Shaping The Battlefield III
Rule 412 Motions
in Military Practice

JOCELYN C. STEWART, ESQ.

Shaping The Battlefield III
Rule 412 Motions in Military Practice

Copyright © 2022 Jocelyn C. Stewart
Book produced by: Anspach Media
Cover design by: Freddy Solis
Photo credit: Misha Dumov, Tacoma Headshots
Makeup credit: Kaitlin Vigil, Tacoma Makeup Artist

ISBN 13: 978-1-7377355-5-7

Printed in USA

Table of Contents

Disclaimer

The purpose of this book is to provide more specific guidance to drafting and litigating Rule 412 motions. From experience in nearly two decades of litigating sexual assault cases as a military prosecutor, uniformed defense counsel, and civilian defense practitioner, sexual assault cases crowd the docket more than any other single type of case. Rule 412 motions are litigated in nearly each of those cases. Sometimes those motions are well written and litigated, but often they are not. But they always are pivotal.[1]

The target audience is military counsel that are new to military justice practice, for more experienced counsel that may not have had clear guidance provided to them in best practices for Rule 412 motions practice, or for any practitioner who wants to revisit her own process to identify and explore another method.

The advice and directions that follow are based not only on Ms. Stewart's own experience in approaching

---

[1] Two precursor books called "Shaping the Battlefield: How to Draft Motions in Military Practice" and "Shaping the Battlefield II: How to Litigate Motions in Military Practice" were previously published and remain available for purchase in paperback or Kindle format. https://www.shapingthebattlefiled.com

military courts-martial since 2005,[2] but also based on feedback from dozens of military judges who remark about the pitfalls they witness in military motions practice. Additionally, several of these best practices are echoed from conversations with and social media posts and comments from other counsel.

Ms. Stewart has appeared before more than fifty-four Military Judges of various service courts, and she has litigated well over one hundred contested courts-martial since 2005. Ms. Stewart has earned respect, and more importantly, meaningful relief, for her clients through effective and zealous practice. Rule 412 motions can make or break a case, and in this particularized subset of motions there are additional permutations and tactical decisions than are present in most motions.

---

[2] In 2005, Ms. Stewart began work as a new trial counsel where she remained in prosecution litigation for twenty-six months. In 2007, she transitioned into military defense work until 2010. In 2010, Ms. Stewart was selected to serve as one of the original fifteen U.S. Army Special Victim Prosecutors. In 2012, she left active duty to pursue private practice in military courts-martial defense. Ms. Stewart founded her own law firm and is the CEO of the Law Office of Jocelyn C Stewart, Corp. As of 2022 the firm consists of three full time attorneys, five part time attorneys, one appellate counsel, two paralegals, and a dedicated defense investigator. Members of the firm have represented clients of every military service and have defended clients in the U.S., Iraq, Afghanistan, Germany, Italy, Japan, and South Korea. From January 2017 until January 2020, Ms. Stewart also served as adjunct professor of criminal law at The Judge Advocate General's Legal Center and School in Charlottesville, Virginia.

Ms. Stewart believes that motions, and particularly those involving Rule 412 in sexual assault cases, shape the battlefield of trials.[3]

---

[3] There are sound strategic reasons not to file certain motions that are not waived by waiting to use the issue as an objection. If there is a factual predicate that falls under Rule 412, these motions *must* be filed. Policy will dictate that if not provided the requisite notice, the facts that may be critical to your case, may never be known to the finder of fact. This book will not analyze when to file a motion and when not to file a motion. Instead, this book will provide a primer on the best practices of drafting and litigating Rule 412 motions that the attorney is required to file in advance of trial. If you learn later of Rule 412 evidence, you will need to ensure the military judge finds "good cause" for your late filing.

Foreword

Sean F. Mangan, Lieutenant Colonel (Retired), U.S. Army

This book is the third in Ms. Stewart's excellent **"Shaping the Battlefield"** series of how-to guides for courts-martial litigation. It follows her two earlier books on drafting and litigating motions in military trials. Unlike previous books, this version steps forward into a specific area of litigation: the complex and challenging realm governed by Military Rule of Evidence (MRE) 412. With its heavy emphasis on sexual assault-type crimes, the military litigates "412 matters" more frequently than other jurisdictions. This makes MRE 412 required reading for any military justice practitioner, uniformed or civilian. Whether something "is or is not MRE 412" is a question frequently asked. The rule's ambit reaches far beyond trial, impacting administrative boards, Article 32 preliminary hearings, and many other actions. While this book is focused on motions practice, it also can serve as a valuable resource outside courts-martial as well.

A confluence of law and policy, MRE 412 was the rule most discussed in the Criminal Law Department during my years teaching at The Judge Advocate General's Legal Center and School. The product of interplay between modern policy objectives and established legal principles,

15

MRE 412 is not unique because of its content. The legal concepts involved—character, privacy, relevance, weight, source of injury, probative value, constitutional necessity—all are longstanding and well-developed. What sets MRE 412 apart is how it came to be. Reaching into America's courtrooms in a way not often done, Congress deliberately altered what evidence may be used for certain allegations. First appearing in 1978 as a federal rule limited to criminal trials for the crime of rape, Federal Rule of Evidence (FRE) 412 was expanded in 1988 into a broader rule covering all sex offenses, and then adjusted again in 1994 to apply generally as a federal evidentiary rule. As with all new FREs, Rule 412 was adopted and adapted into the military court system as MRE 412. This was a significant reversal of military rules on the subject: previously, paragraph 153 b (2)(b), from the 1969 Manual for Courts-Martial allowed "any evidence, otherwise competent, tending to show the unchaste character of the alleged victim."

In the decades that followed MRE 412's creation, the U.S. military saw a major shift in awareness and attitudes about sexual crimes. The subject was advanced by larger social movements of gender equality, victim's rights, and public accountability of the military. Inside the military, the issue was forced into the light by changes to the Armed

Forces' composition and deliberate efforts to remake military culture surrounding sexual harassment and sex offenses. This momentum continues in the current national discussion, most recently realized in the #MeToo movement that dominated social and news media. Facing congressionally levelled accusations of a "rape culture," and threatened with the loss of its service-controlled criminal court system, the Department of Defense took action to combat sex-related crime in the ranks. Among solutions considered was amending MRE 412 to secure more convictions in courts-martial trials. No other rule of evidence in military jurisprudence has had such attention, and nothing suggests that the debates and developments surrounding this rule will end any time soon.

For the practitioner, MRE 412 brings law, procedure, and evidence together in a challenging way. To succeed in MRE 412 litigation, an attorney must prevail across the full range of legal skills: identifying and seeking out facts in evidence through review and investigation, establishing those facts on the record through exhibits and witnesses, identifying a legal theory that ties those facts to authority, developing persuasive arguments that advances that theory over competing interests, and finally communicating those arguments orally and in writing to a Military Judge. In addition, the large and diverse body of military caselaw surrounding MRE 412 demands careful research and analysis from the trial lawyer.

MRE 412 also gives supervisors a valuable tool to evaluate and develop military counsel. As a regional supervisor in the Army Trial Defense Service (TDS), I found that I could learn more about the talent and skills of each of my 30 defense counsel through one MRE 412 motion than I could watching some entire trials. I wish I had appreciated this when I was a Chief of Military Justice years before, as I believe MRE 412 provides a great opportunity to assess subordinate counsel regardless of whether they prosecute, represent victims, or defend. I even recall the watchful gaze of the Army Chief Trial Judge when he observed me presiding over my first MRE 412-related trial. Challenging and complex, MRE 412 is a rule that demands much but teaches even more.

In the pages that follow, Ms. Stewart takes MRE 412 apart. Like the two volumes that precede it, this "Shaping the Battlefield" guide walks the reader through litigation step-by-step, making principles accessible and giving useful advice. Because her perspective benefits from nearly two decades of continuous military litigation, Ms. Stewart's guide to navigating MRE 412 is a valuable resource for all, regardless of party or role. It is a subject worthy of attention, because MRE 412 is also something else: decisive. Winning or losing a MRE 412 motion can mean winning or losing the entire court-martial. This book fills an important gap in existing legal training and underscores the significance of MRE 412 in military criminal practice.

Jocelyn C. Stewart

Introduction

Jocelyn C. Stewart

As I shared in the precursors to this book, motions are a pivotal step to shape the battlefield of courtroom practice. This is particularly true in litigating Rule 412 motions in advance of sexual assault trials.

In recent history, the bulk of courts-martial that are being tried in military courtrooms allege a form of sexual assault under Article 120, UCMJ. Unlike other types of motions where there can be strategic elements of deciding whether to file the motion at all, these motions *must be filed.* There are additional notice requirements that are particular to Rule 412 motions, and without filing with the proper notice, you risk that your evidence will not be deemed admissible.

The nuance of Rule 412 litigation is not in whether to file, but in how to file and most significantly in connecting the dots to ensure success so the judge will rule for inclusion. You will often hear judges and practitioners harkening back to the idea that Rule 412 represents a rule of exclusion, vice a rule of inclusion. Your role as the advocate is to articulate a basis for the inclusion of the facts when battling against the uphill presumption of exclusion.

Over the last decade or so, military practice has seen a significant ebb and flow of cases involving prior sexual behavior of alleged victims. Perhaps none was as significant as *United States versus Ellerbrock*. [https://bit.ly/USvsEllerbrock]. I served a special victim prosecutor from 2010 to 2012, and I was chosen to head up the *Ellerbrock* retrial. Based on case milestones and a significant last-minute defense obtained continuance, I did not have the opportunity to try the case before leaving active duty. I lament the result in the case as someone who convinced the alleged victim to put herself through a second trial. The *Ellerbrock* opinion, which gave rise to the retrial, no matter how I feel about its conclusion, has been monumental in assisting defense counsel in articulating other acts evidence as relevant to a motive to fabricate.

Another significant moment in Rule 412 practice was Congress' attempt at "removal" of the constitutional exception. As many practitioners surmised, this congressional change did little to bring about practical impact. The military accused's constitutional protections remain concretely in place.

From anecdotes and questions that I have received while in private practice, I know that motions practice and particularly that under Rule 412, needs attention in military practice. For example, in Hawaii a Military Judge excluded an exclamation of an alleging victim who declared to party goers that she wanted sex the night of her alleged assault (though her language was not so tactful). The accused was present when she made her public desires known. I had a hard time understanding the rationale when I learned that the judge excluded the evidence. After additional probing and reading the judge's findings of fact and conclusions of law, I uncovered the defense counsel did not provide any evidence to the Court that the accused's belief that the alleging victim wanted to engage in sexual activity with him during the charged offenses had been impacted by having heard the public statement an hour or two earlier in the night. I gently nudged the defense counsel to file a motion for reconsideration with an affidavit from their client and prompted this revision.

Rule 412 litigation is not subtle. Judges will not connect the dots for counsel. If you are uncomfortable talking about the details, then chances are you are not doing your job. Shed your shame and do your job for your client.

Recall that it is a rule of exclusion, and the Court will not rule evidence admissible when counsel have not laid out relevance and materiality in a painstaking and clear process.

The "cut and paste" approach by defense counsel to the drafting of Rule 412 pleadings in military motions practice is causing a demonstrable challenge to successfully defending sexual assault allegations. I have been unable to discern whether the disconnect on Rule 412 litigation is caused by the broader deficit in motions training, from a desire by counsel to somehow keep their cards close to the vest, or from some other dearth.

I am heartened that more recently training officers in the Army's defense bar are renewing emphasis on motions practice. The import of this particular type of litigation cannot be underestimated. Failing to approach Rule 412 motions practice as a pivotal step in shaping the battlefield for the war of a sexual assault court-martial trial can cost a military accused his liberty and render him a registered sex offender.

This book dares you to step away from the "shared drive" and instead to take a fresh approach with each motion for every case. I have dissected my own process and am providing the insight that I derived for your benefit in the form of this book. My law firm's[4] mantra is that we want to be where we can do the "most good."[5]

My non-profit's[6] goal is to provide training so that we might collectively "Raise the Bar" in military court-martial defense practice. My goal for this book is the same: to elevate the practice of sexual assault courts-martial in the military. One Rule 412 motion at a time.

- Jocelyn C. Stewart -

* In 2012, Ms. Stewart left active service to pursue a private practice dedicated to defending servicemembers worldwide, with particular emphasis in sexual assault investigations and sexual assault court-martial defense. As a civilian court-martial and military administrative matter specialist, she has represented clients OCONUS and all over CONUS, from the west coast to the east coast, the gulf coast, and the Midwest. Her clients serve the Army, the Air Force, the Navy, the Marine Corps, and even the Coast Guard.

---

[4] Law Office of Jocelyn C Stewart, Corp. www.ucmj-defender.com

[5] https://www.ucmj-defender.com/we-want-to-be-where-we-can-do-the-most-good-for-military-justice/

[6] International Association of Military Defense Lawyers (IAMDL) www.iamdl.org

Jocelyn C. Stewart

PART I:

DRAFTING RULE

412 MOTIONS

Jocelyn C. Stewart

Chapter 1

WHAT IS A RULE 412 MOTION?

WHAT IS A MOTION?

Recall that a motion is a written request for relief. Your starting point for any motions writing endeavor is Rule for Court-Martial [RCM] 905.

RCM 905(a) defines a motion and explains the form of a motion. Its discussion provides the groundwork that there are four types of motions: "Motions may be motions to suppress [(see R.C.M. 905(b)(3))]; motions for appropriate relief (see R.C.M. 906); motions to dismiss (see R.C.M. 907); or motions for findings of not guilty (see R.C.M. 917).

WHAT KIND OF MOTION IS A RULE 412 MOTION?

A Rule 412 motion is in the category of motions for appropriate relief. You are asking the Court for a specific type of relief that permits you to introduce certain evidence that falls under Rule 412.

WHAT IS A RULE 412 MOTION?

Specifically, Rule 412 motions are a party's written request to the Court for a finding that matters that would otherwise be precluded from admission about an alleged victim's prior sexual history or predisposition has an admissible purpose. In the first part of the rule, Rule 412 generally precludes evidence of an alleged victim's sexual history or predisposition. However, the subparts that follow the initial rule outline exceptions to this general rule of preclusion.

Before we dive into the exceptions to the rule, one of the most important components of Rule 412 is that the rule has strict notice provisions. There are three components of Rule 412's notice provision: first, the rule requires *written* notice; second, the rule demands that the noticing party provide the *specific purpose* for which the evidence is being offered; and third, the written notice must be provided *at least 5 days prior* to the entry of pleas.[7]

It is important to note that while the majority of pretrial litigation pursuant to Rule 412 is initiated by the defense, the rule applies equally to the government.

---

[7] Rule 412 (c)(1) reads in pertinent part: A party intending to offer evidence under subdivision (b) must—
(A) file a written motion at least 5 days prior to entry of pleas specifically describing the evidence and stating the purpose for which it is offered unless the military judge, for good cause shown, requires a different time for filing or permits filing during trial; and
(B) serve the motion on the opposing party and the military judge and notify the victim or, when appropriate, the victim's guardian or representative.

There are occasions when the government may wish to introduce evidence under Rule 412 that supports their theory of the case. One example could be if the alleged victim has a prior sexual trauma which is relevant to why the alleged victim responded to the current alleged event in the way that he or she did. If there was evidence of "freeze" rather than flight or fight, for example. If the government wants to introduce mention of the prior sexual trauma, the government must abide by the same notice provisions (and pleadings) as the defense.

Because Rule 412 requires specific notice of the purpose for which the evidence is being offered, a Rule 412 motion must include the exception under which the evidence is being offered. In addition to the exception for which the evidence is being offered, as I will discuss later in Chapter 4, it is also critical to spell out how the evidence supports the theory of the case in some material way.

Chapter 2

HOW TO LOOK FOR RULE 412 FACTS FROM THE CLIENT

In every Article 120 case, you must identify if there are facts that you would need to introduce that potentially fall under Rule 412. While true for both the government and the defense, I believe it will be most critical for the defense.

USE THE CLIENT

The first step that I use to identify facts for potential Rule 412 litigation is to begin with the client, the accused person. Remember that in all litigation, there is an inherent advantage to the defense counsel: access to their client and his knowledge of the case. Except in cases where the client cannot recall many facts due to alcohol or other incapacitation or professes to not have had any contact at all with the alleging victim, the client is a resource to which you have unfettered access that the prosecution does not. This is particularly true when the client invoked his right to counsel and / or silence and has not made any pretrial statements.

MISTAKE OF FACT AS TO CONSENT

In most defenses of sexual assault allegations, the critical issue for the panel will be whether or not the military accused believed his partner was consenting at the time of the conduct. This is known as mistake of fact as to consent and so long as the belief was a reasonable one, represents a complete defense to all forms of sexual assault of people over the age of 16 years.

HOW TO ELICIT WHAT YOU NEED

In my practice I have identified several ways of streamlining defense work and Rule 412 litigation is no different. One of the ways that I make this work more efficient is to assign a "homework" task to my client that covers this litigation. Please note that I do not tell the client the purpose for this "assignment." I merely assign homework to my client.

I begin by giving this exercise to the client at a distance for several reasons. First, it will give him time to consider his thought processes; this can often be clearer than when you are peppering him with questions. Second, it may enable the client to feel more comfortable by writing it down first before being asked about it by someone who he is still learning to trust.

Realize that in Article 120, UCMJ allegations the mistake of fact defense has two components – you must first identify the client's personal knowledge and beliefs, which comes from his observations and the perceptions.

ASSIGNING THE ACTUAL HOMEWORK

I tell my client to list out every reason or event *leading up to the contact* that caused he (or she) to believe or understand that the alleging victim wanted consensual activity. I also tell my client to list out every indication (noise, action, word, etc.) he or she received from the alleging victim *during their sexual encounter* that he or she was actively engaging in or wanted to engage in sexual contact with the client. Lastly, I invite the client to inform me about every behavior, communication, or action *after the sexual encounter* that reinforced for them that their intimate partner had consented to the sexual events.

USING WHAT THE CLIENT GIVES YOU

Not every fact or piece of information the client provides to the lawyer as a part of the above described exercise will provide evidence of mistake of fact for Rule 412 purposes. Some of the behaviors will not even be sexual or constitute predisposition. Pay particular attention to what the client later learned. Understand that what a client did not know at the time of the sexual encounter cannot impact whether or not the alleging victim was consenting or whether he believed she was consenting because he did not know it at the time of the contact. [This is not to say that information later learned does not have another purpose, but it will not matter to this exercise.]

The portion of your assignment about what the client knew before the events and what the client observed during the events are geared to consent as a defense or mistake of fact as to consent as a defense. This is the first step in screening for Rule 412 facts.

NON-SEXUAL FACTS

Not every fact on the client's list will be sexual predisposition or sexual history. For example, a client may place on the list the fact that the alleging victim wrote her phone number or room number on the client's hand with a pen. Unless the writing included something expressly sexual, this fact may not need to be decided through Rule 412 litigation. Perhaps the alleging victim placed her hand on the arm of the client in a familiar way and smiled at him. This is not an example of sexual predisposition or behavior. But the behavior is relevant to your client's mistake of fact and gives you an important detail to elicit in some way during trial.

This exercise is also important to help refocus your client on what he knew and experienced prior to and during the sexual encounter. Realize that often after an allegation surfaces, your client may receive additional information from his supporters about the sexual history of the alleging victim. What your client later learns may or may not be relevant, but information later gleaned will never be relevant for a client's mistake of fact at the time of a sexual encounter. Perhaps the information is useful to developing evidence of bias or motive to fabricate, but even if admissible information about the putative victim that the client did not know at the time of the sexual encounter will not be admissible for this aspect of Rule 412. Please see the discussion in Chapter 3 about how to develop additional Rule 412 admissible evidence.

ACTUAL HOMEWORK

There is a significant advantage to requiring the client to type out their lists. Providing a list in writing helps you in many ways. First, it gives you a starting off point to screen for facts to include in a Rule 412 motion. Second, it may make it easier for your client to become accustomed to discussing intimate details of the encounter and thereby build rapport and trust with you as the attorney. The exercise I describe of having the client outline in writing and list those factors and events that he experienced will help you in forging the trusted attorney-client relationship. This process is the first step of their potential preparation to take the stand in their own defense. Third, after you conduct an interview with the client, all of the facts will provide you with a checklist for the next phase of your analysis.

OVERALL SCREENING

Not every fact the client provides in their lists may be a fact you wish to assert at your trial. There are certain facts that may not support your case theory.

Likewise, the client may assert facts that even if true and supported by evidence would be precluded from the Court as a matter of course. An important component of Rule 412 litigation is learning to act as your own gatekeeper. Part of appropriate gatekeeping is not offering evidence you know has no lawful purpose.

Jocelyn C. Stewart

Chapter 3

HOW TO LOOK FOR RULE 412 FACTS FROM OTHER SOURCES

The first place to identify facts for potential Rule 412 litigation is the client because what occurred in almost all cases also happened to him. However, realize that he is not the only potential source of Rule 412 facts.

Once you digest the written "homework" of the client, you will need to turn to other methods and sources that will hopefully independently corroborate the information the client provided to you. If the client has an absence of memory, these other secondary sources become even more vital. In this way, your litigation may focus on contradicting the details surrounding the alleging victim's allegation. Remember that all cases are unique.

Understand that even if there are no additional witnesses to the sexual encounter, the corroboration or lack of it will focus on the before and the after of the encounter. This process looks to the potential biases and potential motivation to fabricate. Potential bias also examines whether there are behavioral health components to the alleged victim's experience.

In examining biases and potential motivations to fabricate, one area to try to cultivate Rule 412 facts is to examine the *timing* of the outcry. Did it come when the alleging victim was herself facing potential UCMJ or administrative separation? Or did it potentially coincide with when rumors of the alleging victim's promiscuity became open and notorious?

MEMBERS OF THE AV'S TEAM

The best way to cultivate these potential facts is to speak to the members of the same team, squad, and / or section of the alleging victim, and for supervisors who can identify who those people may be. These are people to whom the complainant may have reported, said something derogatory about your client, or to whom she confided other personal intimate information.

Even though this preliminary screening and interview process is for the purposes of potential motions, each one of these interactions is an opportunity for overall trial strategy. While searching for Rule 412 evidence, each of these people should be screened about whether the complainant provided them any information about the alleged assault or said anything to them about your client or even about their expectations about the upcoming litigation or case. In more than one case, I have seen alleging victims brag about how "their lawyer" told them that the accused was "screwed." This callous attitude does not play well to the panel.

TRUSTED INDIVIDUALS

Another area to potentially develop Rule 412 facts is to attempt to interview people who are close to the alleging victim. Often the alleging victim may have confided in trusted individuals about the encounter. Perhaps those versions of what she said to different people have substantial differences. Perhaps the complainant confided in her trusted friends that this sexual encounter was reminiscent of a prior trauma. In fact, we see it repeat itself in cases where a person was reportedly the victim of multiple different sexual traumas over their life and even starting from childhood.

Note that accepting as true the prior trauma is far more likely to be admissible than to attack the alleging victim as someone who makes false allegations of sexual assault. Hint: the latter argument is propensity evidence, and the same rules apply to you as would the government.

However, there is support in military case law to explore admission of al alleging victim's prior trauma of and how those prior events can impact a person's perceptions about a given (charged) event.[8]

---

[8] According to the Court of Appeals for the Armed Forces in their 1998 opinion in United States v. Soifer, 47 M.J. 425:

FAMILY MEMBERS

Realize that family members may or may not also be a potential for cultivating Rule 412 facts. Often counsel do not try to reach out to family members, assuming that the family members will not speak to the defense. I would caution you against foreclosing on this potential avenue without even trying. You just may be surprised.

In a recent case with a several year delayed report, the alleging victim told law enforcement that the only people that she told about her experience were her parents. Her father revealed that he had no idea about any claimed sexual assault and that his daughter had never shared anything about a prior attack.

How a witness "views" an event, in terms of her five senses, depends on her background, including family life, education, and day-to-day experiences. Witnesses "behave according to what [they] bring to the occasion, and what each of [them] brings to the occasion is more or less unique." In that sense, each witness has a bias. Additionally, a witness's interpretation of an event depends on whether her perception is impaired… A past or present mental condition also may impact on a person's ability to perceive. Id at 428.

The Soifer Court clarifies that potential bias is more than a person's intentional motivation to fabricate an allegation; rather, bias represents the totality of those experiences that can converge to impact how a person interprets an individual experience.

In another case, the spouse of the alleging victim submitted to an interview of more than two hours. His interview provided ample evidence that the alleging victim had used makeup to fabricate bruises on her neck for a FaceTime call with her spouse. None of these bruises were present when she reported for her sexual assault forensic exam hours before the call. Other details she gave him, and his responses to her completely contradicted the version she gave law enforcement. His interview also led to the development of substantial independent evidence that we never would have known to explore but for this defense interview.

Note that the law enforcement report annotated two lines from their phone call to this spouse. The report attempted to make it seem as though the alleging victim's call to the spouse was an emergent cry for help when in fact it was a routine scheduled call two days later. Never trust a law enforcement report or take it at face value. There is always more to mine.

RUMORS

I would caution you about rumors. Unless they can be traced to a given person who at least "heard" about the Rule 412 fact, you likely will not reach the threshold preponderance standard. However, if you can provide evidence that the alleging victim was aware the rumors were circulating, the truth or veracity of the underlying rumor is not what you are attempting to introduce. There will be more to follow in subsequent sections about really drilling down on what it is you should be asking to introduce. For these purposes, realize that rumors need to either be substantiated to reveal admissible facts or you will need to substantiate the alleging victim's awareness of rumors as a potential motivation to fabricate the assault (to deflect from rumors of promiscuity, for example).

Investigation and thorough exploration of potential avenues of Rule 412 facts are critical before any writing or narrowing of defense theories can occur.

Jocelyn C. Stewart

Chapter 4

RETURN TO THE RULE

RETURN TO THE RULE

Now that you have conducted investigation and outlined a number of facts that are potentially admissible regarding an alleged victim, you must turn to the <u>rule</u> itself.

Recall that Rule 412 sets out at a general *prohibition* to admitting evidence of an alleged victim's past sexual history or sexual predisposition.

Many practitioners jump to the exceptions to the rule. Before even turning to the exceptions, I find it important to remind practitioners that not every sexual fact about an alleged victim is even subject to an exception. What do I mean? Look to the text of the rule itself. In the Venn diagram of sexual evidence, realize that there is a substantial zeroing in contained in the rule's general prohibition. This rule governs *(1) Evidence offered to prove that a victim engaged in other sexual behavior; or (2) Evidence offered to prove a victim's sexual predisposition.*

Ask yourself: is there an argument to be made that the evidence you are offering is not actually prohibited by Rule 412? Does this evidence trigger the prohibitive provisions of the rule? Sometimes that means diving deeper on what it is that you actually need to introduce at trial. Sometimes this means asking yourself: what is the purpose for why you want to offer it?

WHY ARE YOU OFFERING IT?

First, ask yourself, why are you offering the evidence. Are you offering the evidence to prove that a prior sexual event occurred? Are you offering the evidence to suggest that the alleging victim has a specific sexual predisposition? Does it really matter that the alleged victim engaged in a prior sexual event? If the conduct or behavior was with your client or directed toward your client, then the answer is probably, yes. Does it really matter that the alleged victim actually has a sexual predisposition toward being promiscuous? Or does what matters is that the alleging victim was concerned she was developing a *reputation* for having a specific predisposition toward sexual activity?

Largely, I believe that we as practitioners (and even some judges, at times) as guardians lose sight of the precise language of the rule. I tend to think the rule itself needs to be revised. If the evidence is being offered to prove sexual predisposition, then it should never be allowed. Yes, you read that correctly. Offering evidence to prove an alleged victim is promiscuous means that you are arguing the fundamental prohibition and purpose for the rule itself – you would be arguing that the alleged victim was not assaulted on a specific occasion because she would never say no. That is what Rape Shield laws were enacted to prevent.

If you are not offering the evidence to provide the underlying behavior, then argue that Rule 412 does not govern your motion. File the motion out of an abundance of caution, but your first argument should be that this evidence is not prohibited because you are not offering the evidence *to prove that the alleging victim had a specific sexual predisposition*, but rather the impact of a person's knowledge and attendant belief about behaviors.

Look to the Rule. Much of what litigants argue during Rule 412 litigation is not actual Rule 412 evidence. Any opportunity you have to help the judge avoid the Rule 412 analysis is one step closer you are to ensuring inclusion of your desired evidence.

Chapter 5

HOW TO IDENTIFY THE RULE 412 BASIS FOR ADMISSION

Some evidence is Rule 412 evidence and requires its litigation. After conducting a meaningful analysis about what does qualify for litigation, turn to the exceptions.

SOURCE OF SEMEN, INJURY, OR OTHER PHYSICAL EVIDENCE EXCEPTION

Most practitioners will not see this exception during their practice. Litigating these matters when it does arise is fairly straight forward. The most common of this exception is when a DNA report notes a mixture of contributors to the minor profile (i.e., who is not the alleging victim), and it is more straightforward when the source of the DNA can be identified as semen.

In addition to multiple semen contributors, I have seen this exception arise in a case involving an allegation of forcible rape by strangulation. There were noted injuries to the side of the neck, and the alleging victim attributed these injuries to the conduct of my client. In conducting investigation into the alleging victim, we learned that she had recounted to a close friend that she had been hospitalized from a different intimate partner who had been routinely strangling her during sex.

While rare, evidence of an alternative source of semen or injury is extremely powerful evidence. Not only does its existence assist military clients in obtaining relief in Rule 412 litigation, but also, we see it have dynamic impacts in the verdicts we see.

PRIOR SEXUAL ACTIVITY WITH THE ACCUSED PERSON EXCEPTION

When the accused has had a sexual history with the alleged victim, this evidence also can be powerful. Whether the two were married or merely "hooked up" on one night, prior intimacy matters. What could be more probative about whether the accused person believed his partner was consenting?

Some judges view different levels of intimacy as triggering Rule 412 litigation. Early in the military's emphasis on prosecution of sexual assault, Military Judges were far more inclined to only apply Rule 412 to intercourse. In contemporary litigation, Military Judges lean the other direction and tend to apply Rule 412 to flirtatious behavior and intimate contact far short of intercourse.

Practitioners are better off pleading any form of intimacy to shape their battlefield. You would hate for a gametime call to go against inclusion because the judge did not appreciate the lack of notice and litigation. Remember, help the judge help you. Judges are more apt to make decisions that benefit inclusion of evidence if they have been briefed and given a fair opportunity to weigh and evaluate all pertinent factors.

CONSTITUTIONAL EXCEPTION

Like many rules in the law, there are exceptions. Rule 412 outlines what are its specific exceptions. In one iteration of Rule 412, law makers attempted to "delete" the constitutional exception contained in Rule 412. Most practitioners agree this is absurd. The constitutional right to present a complete defense, within appropriate limits, should give way to the risk of embarrassment to an alleged victim.

One of the most effective ways to help provide solid constitutional basis is to be reasonable in your prayers for relief. Remember that the more detail you attempt to admit into evidence, the less likely the judge will be to grant your motion.

In the world of Rule 412 litigation, the more intimate details the more likely you are to trigger the balancing test. And not in a good way. Ask yourself, what is the minimum of what I actually need to admit into evidence to be able to make the arguments I need to make. Chances are you do not need any detail about intimacy with someone other than the accused. The relationship itself is far more likely to be admitted under a constitutional exception and there is no need to bring up the graphic details to do it. Keep it classy, and always constitutional.

Jocelyn C. Stewart

Chapter 6

PROVIDE A CLEAR REMEDY FROM RULE 412

BE CLEAR

Probably the most frustrating aspect of poorly drafted motions about which judges complain is that the judge has to work to understand what it is the counsel is asking the judge to do. If the judge has to work at comprehending your requested remedy for relief, then it is far less likely that the judge will do what it is you want for your client. So, help the judge help you!

First, decide what it is that you want. Then, figure out what (from the rule) you are asking the judge to permit you to admit. Next, identify the mechanism you want to be able to introduce the evidence. Each of these components is required in Rule 412 notice.

Being clear almost always means being concise. Consider drafting the remedy and then challenging yourself to delete at least 30% of the words from your first attempt. Phrase the desired remedy in a way that tells your judge clearly and concisely what you want him to do. Remember that a critical part of identifying the desired remedy is to provide the mechanism for which the military judge should grant you relief.

IDENTIFY THE SUBPART

Rule 412 contains at least part of the remedy you are requesting. You must identify which subpart the judge should apply to permit your evidence to be introduced. Where practitioners often fall short in the remedy is not identifying how exactly you are asking that the evidence be permitted into evidence. If you do not write it in your motion, the judge will ask you during oral argument, and I promise you that is not an ideal time to be trying to work through it. You may miss a critical mechanism, and if you do not identify it at least by the time of oral argument, you face significant hurdles to try later.

If you are seeking introduction under the constitutional exception, you are also going to want to identify a military appellate case that has previously found this kind of evidence constitutionally required under Rule 412. If you are using the constitutional exception, remember to include the U.S. Constitution, (e.g., the Due Process clause) also in your prayer for relief.

If you do not base your request for relief on the appropriate subpart of the rule, if the judge denies your request, you may waive the issue on appeal.

This is not a matter of emphasizing form over substance, but rather a systematic approach to ensure that your prayer for relief is accurate, covers all grounds, and to protect your client on any potential appellate review.

SAMPLES

Samples of a properly phrased requested remedy in Rule 412 motions are as follows:

IAW [9] Military Rule of Evidence 412(b)(1), defense moves this Honorable Court to permit introduction of the DNA mixture from AV's cervical swab to prove another source of semen or injury. Defense seeks introduction through cross-examination of the government DNA expert if the government calls their expert and it is fairly raised, or through direct examination of the government DNA expert if the defense calls him instead.

Pursuant to M.R.E. 412(b)(2), defense seeks introduction of evidence of the prior intimate relationship of the [Client Rank and Name] and AV to prove consent or mistake of fact as to consent for

---

[9] IAW is an acronym often used for "In accordance with."

the charged offenses. Defense seeks to introduce this evidence by cross-examining AV about the fact that they participated in previous sexual activity, the frequency of the activity, and timing of their prior intimacy after getting into disagreements. If [Client Rank and Name] decides to testify in his own defense, defense also seeks to introduce this evidence during his direct examination.

Defense respectfully asks to introduce evidence of AV's relationship with her husband and her fear of losing that relationship through M.R.E. 412(b)(3), the due process clause to the U.S. Constitution, and United States v. Ellerbrock. Defense seeks to introduce this evidence in cross-examining AV and in calling AV's best friend, SPC Green, during the defense merits case.

State the subpart for the rule. If the subpart is the constitutional exception, cite to the Constitution. If there is a seminal case, include it also.

Give at least a general description of the evidence you want introduced. You can either give a more detailed description in the remedy or you will need to be much more specific in the facts of the motion or in the affidavits or testimony elicited during the motions hearing.

WHAT MECHANISM DO YOU WANT TO USE?

Tell the judge in the remedy whether your desired evidence will come into evidence through cross-examination or through direct examination. List each witness through whom the examination must be conducted. Always tell the judge whether there are any contingencies. Remember that trials are fluid, and you will want to account for the different potential routes that your evidence may need to be adduced. Otherwise, you will potentially box yourself out of being able to present vital evidence. For an example of a contingency please see the above sample remedy involving the DNA mixture evidence.

Chapter 7

IDENTIFY THE RULE 412 EVIDENCE

EVERY MOTION REQUIRES EVIDENCE

The recitation of facts in your motion is not proof of the underlying fact. In every motion, you will need to supply evidence to the judge. That evidence is necessary for the judge to make her findings of fact. That bears repeating: *simply because you include an assertion as a fact in your motion and number it and organize it does not make it so*, and most importantly does not enable the military judge to make a finding of fact in a ruling based on any assertions you put into your motion.

A proffer by counsel, which is a recitation of what the attorney expects a given witness would testify regarding, is _not_ evidence on which the military judge can make his finding of fact.

If you draft the motion as you should nearly every paragraph in your facts section has a corresponding appellate exhibit with evidence to support the factual paragraph. One exception is if you do not have a document for support and instead would need to call a witness to support that particular fact.[10] The other exception is if you

---

[10] Please note that I typically discourage calling non-Rule 412 witnesses during motions hearing. For more on this issue, please read Shaping the Battlefield II: How to Litigate Motions in Military Practice.

obtain an affidavit from your client to support one or more facts in your motion and you do not file the client affidavit until the time of litigation.

RULE 412 MOTIONS REQUIRE SPECIFIC EVIDENCE

You must provide a factual predicate for the underlying evidence that you intend to offer under Rule 412. Do not assume the judge will draw inferences because Rule 412 is a rule of exclusion. The judge will not fill in any blanks for you.

If your motion relates to the complainant's reputation, then you must prove the reputation. Do more than elicit this evidence from your client. Obtain one or more affidavits from a member of the same community (i.e., squad, platoon, company, battalion / squadron).

If your motion relates to the alleged victim's prior sexual history, then you must provide evidence of the (relevant) prior history. Rumor will not suffice. If all you have uncovered during your investigation is rumor, you should reconsider drafting and filing this motion at all. If you are setting out to prove prior sexual history with anyone other than the client / military accused, then obtain an affidavit from another person with knowledge, namely, the other person with whom the alleging victim engaged in the activity or a person who witnessed it. I am not referring to someone who merely has heard it happened unless that person has direct knowledge of the alleged victim's knowledge of the rumor and upset over it.

The only way whether a given person "heard" that the behavior happened is relevant to litigation is if that person is the military defendant accused, AND then only if the client was aware of the rumor prior to engaging in the charged offenses. If this is the case, remain wary about whether or not to file the motion. A savvy prosecutor can argue that your client had heard some rumor and that if it impacted his thought process to engage in sexual activity with the complainant, it could represent evidence of motive, particularly if making that leap in his mind is not a reasonable one.

Let me speak plainly. If your client heard a rumor that the alleging victim was promiscuous and that alone (not behavior directed toward him) caused him to believe she was open to sex with him, then perhaps you are proving a pathology that identifies the client as a sexual offender.

EVIDENCE OF PRIOR SEXUAL RELATIONSHIP WITH THE MILITARY ACCUSED

If the motion's goal is to introduce evidence of a prior sexual history between the defendant and the putative victim, then use an affidavit from the client. But do not stop there. Many attorneys do, and I find it to be a tactical mistake.

In light of the rules that almost always foreclose on the defense's ability to conduct a pretrial interview, why would you pass up an opportunity to elicit on the record evidence from the alleging victim? Call the alleging victim to the stand.

EVIDENCE REGARDING MISTAKE OF FACT OF CONSENT AND CONSENT YOU NEED IN A RULE 412 MOTION

If the Rule 412 motion you are filing relates to consent or mistake of fact as to consent, you will need evidence that the evidence impacted your client's belief about why the alleging victim was consenting or why he thought she was consenting. This is the number one piece of evidence that is needed but omitted from Rule 412 motions.

How can the judge possibly find the evidence relevant if you do not relate the evidence back to your client's state of mind?

The best way to provide the Court with evidence of your client's state of mind is to help the client with an affidavit to the Court. And by help, I mean you draft it, and the client verifies it is accurate. If you follow my advice in this book about getting the client to do his "homework" the work on the affidavit is already at least started for you.

Chapter 8

IDENTIFY THE RULE 412 FACTS

WHAT FACTS SUPPORT THE EVIDENCE YOU INTEND TO INTRODUCE?

As with any motion, identifying what facts are needed in order to help the judge reach the conclusion of allowing admission of the evidence is often the most difficult process. After you identify the evidence, you need to prove the underlying conduct or predisposition you are trying to introduce, you need to earmark what is the background that supports it.

FIRST FACTS IN EVERY MOTION

Rule 412 evidence is only relevant in cases involving an allegation of a sexual nature. *See* Rule 412. In every motion that seeks introduction of Rule 412 evidence, the first facts to include in every motion is the nature of the charged offense that is a violation of Article 120, Article 120b, Article 120c, or perhaps Article 128, assault with intent to commit sexual assault or with intent to commit sexual assault of a child. By using the lessons learned from Shaping the Battlefield: How to Draft Motions in Military Practice, you may recall that you are as ever moving from the general to the specific.

MORE SPECIFIC FACTS

After identifying the qualifying charged offense or offenses that enable you to use Rule 412 evidence, you need to identify the more specific facts.

As yourself, "what facts help me lay the foundation or groundwork for the Rule 412 evidence?"

If the motion relates to another source of semen or injury, you need to lay the foundational facts relate to the DNA results: the existence and collection of a cervical swab during a sexual assault forensic examination that the alleging victim experienced.

If the motion relates to the prior sexual history of the complainant and the military accused, perhaps you need to lay the foundation for when and how they met, the status of the relationship and when and how it changed. In this way you are setting the scene.

CONTEXT: THE WHY

The foundational facts in a Rule 412 motion also need to help the judge understand the context for why the evidence you seek to admit relates to your theory of the case.

Rule 412 motions can assist prosecutors in combatting the dilemma of comparative trauma. By the dilemma of comparative trauma, I mean that if an alleging victim's account of what happened is freezing when she was groped momentarily, your panel may not believe her because the charged offense does not seem "that bad." If your complaining witness has experienced a prior trauma and the charged offense retraumatizes her or triggers her, it is easier to explain a freeze by contextualizing this experience for your victim.

The context is a psychological explanation for your victim's behavior. Essentially, a prosecutor would be attempting to corroborate the victim's account for what she says happened and also combat comparative trauma: why it was "reasonable" for your complainant to freeze under the charged set of circumstances. Is it true that people who have not been victims of prior trauma can freeze in a sexual assault? Of course. But if your alleging victim went through this experience before, then your Rule 412 motion may assist you in convincing the panel why they should believe her about the charged misconduct.

If you are a defense attorney, the foundational facts give the context as to why this evidence relates to your defense theory of the case. This seems easier to grasp when the motion relates to prior sexual relations between the complaining witness and the client. Consent and / or mistake of fact as to consent is an express and complete defense. This may seem harder a concept in other Rule 412 motions.

In a motion related to alternate source of semen or injury, the context or the "why" foundational facts also relate to how that evidence supports the defense theory. The theory is either that the alleging victim is confused, cannot accurately recall events, or is lying. How does the presence of another source of semen relate to your theory? Are you espousing that the alleging victim was intoxicated and does not recall consenting to more than one sexual partner? Are you using a theory that the putative victim is blaming your client for her experience because she blacked out to a different later partner? Does the blackout evidence challenge the reliability of her report? There must be a why; you must give context to how the evidence you seek to elicit relates to your defense. If you merely recite talismanic words like "alternate source of semen or injury" without relating its relevance to the defense theory, you are doomed to fail in your motion.

Chapter 9

FACTS, FACTS, FACTS IN RULE 412 MOTIONS

Spend the majority of your time drafting and editing the facts section to your motion. Best put to me early in my career was advice from Colonel (Retired) James L. Pohl who said, "I know the law. I need you to provide me with the facts." Recall that the recitation of facts in your motion is not proof of the underlying fact. That bears repeating. *Simply because you conducted your investigation and learned it from an interview or another source and include those assertions as facts in your motion does mean you are finished.* This is because without including evidence in the motion or during litigation, the military judge will be unable to make a finding of fact to support the ruling you want her to make. A factual written assertion in a motion or a verbal proffer by counsel during argument, is *not* evidence on which the military judge can make findings of fact.[11]

Every paragraph in your facts section should be verifiable from a written document you reference in your motion as a marked appellate exhibit, by written affidavit provided to the court as a marked appellate exhibit or provided at the time of litigation on the motion,[12] or from

---

[11] *See* R.C.M. 905(c), which states, "Unless otherwise provided in this Manual, the burden of proof on any factual issue the resolution of which is necessary to decide a motion shall be by a preponderance of the evidence."

[12] *See* R.C.M. 905(h): "Written motions may be submitted to the military

what will be elicited from a witness during sworn testimony during the Article 39(a) session to litigate the motion.[13]

Writing the facts section to any motion is an evolving process. You likely will find yourself coming back to the facts section to insert additional facts that become apparent you need to incorporate once you have moved onto legal research and composing the legal analysis section. Identifying the necessary facts is probably the most difficult task for attorneys that are new (and for some that are not so new) to motions writing. As you are writing your motion, you must continue to ask yourself what facts the military judge needs in order to grant your desired remedy.

---

judge after referral and when appropriate they may be supported by affidavits, with service and opportunity to reply to the opposing party." I believe that affidavits are under-utilized in motions practice. Often there may be a tactical consideration why you prefer to confine the attestation to those matters set out in the affidavit, rather than automatically guaranteeing the witness will testify during the Article 39(a) and be subject to cross-examination. *See* R.C.M. 912(i)(2) for a definition of witness that includes "one who testifies at a court-martial and anyone whose declaration is received in evidence for any purpose, including *written declarations made by affidavit* or otherwise" (emphasis added).

[13] I will cover in greater detail the considerations of which option to choose from a tactical consideration in my Book "How to Litigate Motions in Military Practice." For now, I will offer that you must decide if it is best to interview the witness before filing a motion and before making your decision which of your options to choose.

THE START OF FACT WRITING

Even in Rule 412 motions, in the first paragraph or in the last paragraph of the preparatory or background facts provide a brief recitation of the allegations / offenses from the charge sheet that are relevant to the motion being drafted. Then cite to the charge sheet without the need to attach a copy of the charge sheet. Repeatedly, I see counsel citing to a document that is already going to be included in the record of trial; the charge sheet is chief among them. Your court reporter will appreciate not needlessly duplicating documents; remember that it always behooves you to keep your court reporter happy.

If the motion you are writing involves the entirety of the allegations (such as a motion regarding an unreasonable multiplication of charges), then at minimum there should be a summary of the entire charge sheet.

If the drafted motion involves only one or two of multiple additional offenses, include the term "*inter alia*", the Latin for among others; note the italics for the Latin term. For example, in a motion where you seek to introduce Rule 412 evidence the motion relates only to one or more of the offenses charged under Article 120, the initial sentence to the first paragraph of the facts section should read:

> [Client / Accused's name] is charged with, *inter alia,* sexual assault pursuant to Article 120.

Emphasizing the one offense at issue signals to the Court that even though the client or military accused may be facing additional offenses, for the purpose of the motion before the judge, the only one that bears the narrow attention is the offense highlighted in the first paragraph of the facts section. In a Rule 412 motion, the sexual offenses are the ones that matter. Often the government charges the same event under differing theories of Article 120, UCMJ; perhaps the Rule 412 evidence only relates to one of them or perhaps it applies to all of them. Make sure you are thinking through each potential way the evidence is relevant.

If the charged offenses also consist of abusing an intimate partner and if the charge sheet includes "sexual offenses" as defined under Rule 412, consider also whether the evidence you seek to introduce is also relevant to that abuse.

After framing the first few paragraphs, you should be ready to conduct research that will best inform the Court about why the judge should grant the relief being requested.

FACTS WRITING IN RULE 412 MOTIONS

Rule 412 motions are non-procedural motions. They are evidentiary motions, which means their facts sections often require more substance. Drafting the facts section of a Rule 412 motion can be a more difficult process. Once you identify the facts you need to include in the manner described in Chapter 10, you must set out to write them.

MECHANICS OF FACT WRITING IN RULE 412 MOTIONS

The following directions guide the majority of Rule 412 motions but should not be construed as the absolute requirement for every Rule 412 motion.

Many practitioners suggest writing the facts section of a motion last. The purpose in writing the facts portion after conducting legal analysis would be to best inform the drafter which facts are the most critical. I concur in part. With regard to Rule 412 motions, the rule itself informs the steps.

Military Judges often complain that practitioners do not clearly outline what precise facts the proponent seeks to admit. I believe this in large part to the litigator's lack of preparation and an absence of organization. As with any motion, there are certain preparatory facts that the judge will need to know in order to understand the Rule 412 facts. By Rule 412 facts, I mean the sexual behavior or predisposition evidence. For example, one preparatory fact in a given Rule 412 motion may be that the complainant and the accused met at a party. That fact alone is not what I refer to as a Rule 412 fact because it is not sexual behavior or predisposition. Another preparatory fact may be that both of them had been drinking alcohol at the time. While not Rule 412 facts, the judge will need to know these facts in order to assess the Rule 412 evidence that you may be seeking to introduce: that the two had intercourse after the initial party where they met and after they had both been drinking alcohol. Perhaps the relevance of these facts is that the prior drunken sexual intercourse between the two impacted the accused's belief that the alleged victim was consenting to sexual behavior on the night of the charged events.

In order to make a distinction between what are the preparatory facts and what Rule 412 facts you seek to introduce, use headers, use font changes, bold characters, underlining, or some other way to signify their discrete separate identity.

I will provide an example for the stylistic approach that I take:

Background Facts:

1. PFC Smith and AV met at a barracks party on or about 31 October 2019.

2. While at the Halloween party, the two were consuming alcohol to the point where it was not advisable for either to drive a car.

3. Not long after the Halloween party PFC Smith and AV began dating exclusively.

4. PFC Smith and AV attended more than 10 barracks parties or parties at off-post residences between November 2019 and January 2020. Both consumed alcohol at each of these gatherings.

5. Sometimes the two met each other at the party and on some occasions the two met at the get together.

6. On 22 January 2020, PFC Smith and AV met at the party in the home of SPC Jones and her wife Mrs. Jones, consumed alcohol at the party, and left the party together in an Uber. The events of what occurred later in the early morning hours represent those offenses that are reflected on the charge sheet. *See* Charge Sheet.

<u>Rule 412 Facts the Defense Seeks to Introduce</u>:

1. During the 31 October 2019 barracks party and while under the influence of alcohol, PFC Smith and AV had sexual intercourse.

2. Between November 2019 and January 2020, PFC Smith and AV had sexual intercourse during or immediately after each gathering that they attended.

3. In the early morning hours of 23 January 2020, when PFC Smith had intercourse with AV, PFC Smith believed she was consenting in part because of their history of engaging in intercourse during or after parties when they had been drinking alcohol.

4. PFC Smith and AV never had sexual intercourse when they were sober and not drinking alcohol.

When you highlight which facts are for background and which facts should be admissible under Rule 412, you help the judge to help you. You are far more likely to receive relief if the judge can tell what specifically the relief is you are seeking.

Jocelyn C. Stewart

Chapter 10

FACTS SECTION ORGANIZATION IN RULE 412 MOTIONS

FACTS ORGANIZATION GENERALLY

Recall that when writing the facts section to any motion that you need to number your paragraphs. Also recall that you should start a new paragraph for each new event or fact. This assists you in being clear and to enable the Military Judge to best follow your motion.[14]

To the extent possible, at the end of each paragraph, list a supporting appellate exhibit for every paragraph that corresponds to the facts averred in that paragraph. Note that when you prepare to draft the evidence / witnesses section of your motion and as you prepare for litigation on the motion during the Article 39(a) session, any paragraph that does not list an appellate exhibit that supports the entire paragraph will require testimony. By separating each paragraph by each changing supporting appellate exhibit, it will also enable you to plan out what witnesses you will need in order to establish each required fact.

By numbering each paragraph and by separating the discrete facts for which you can provide proof, you also enable the government counsel to isolate what facts, if any, with which the government counsel disagrees. When

---

[14] Air Force rules of court dictate that every paragraph of every section of the motion is numbered but the other service courts do not.

government counsel agrees with the defense facts, provided the defense has provided the Military Judge with the facts she needs, litigation becomes streamlined to permit the judge to apply the law to the facts she has been provided.

Change paragraphs at least as early as when you cite to a different piece of evidence to establish the discrete fact or facts. Not every sentence requires isolation into its own paragraph, but some sentences will stand on its own. Please see my previous example of facts in a Rule 412 motion from Chapter 9 of this book.

Isolate paragraphs that require the testimony of a proponent witness during the Article 39(a) session. For maximum clarity, also ensure that you separate facts into separate paragraphs if they require the testimony of separate proponent witnesses, with at least one paragraph per proponent witness.

FACTS ORGANIZATION IN RULE 412 MOTIONS

In Rule 412 motions, the facts should be plainly organized to distinguish between what are predicate or background facts and which facts constitute evidence that falls under Rule 412. The writing process of a Rule 412 motion and its organization may even crystalize for you whether you need to file the motion or whether there are certain predicate or background facts you will need to establish logical and legal relevance.

It is also imperative that your facts organization makes plain which facts you want to admit that fall under the specter of Rule 412. Ensuring facts organization will help the Military Judge identify why a closed session Rule 412 hearing is required, which is part of the analysis he must perform before closing the court. Identifying and separating out the facts in an organized manner also streamlines your plan for litigation.

Chapter 11

CONCLUSION OF RULE 412 MOTIONS

CONCLUSION IN ANY MOTION

Prior to sinking into legal research and analysis, I recommend copying the relief sought into the conclusion of your motion as the framework for the conclusion. Recall that you need not copy the rule's language into the conclusion as you may have done in the prayer for relief.

When concluding a military motion, one needs to circle back to the prayer for relief, a brief *summary* of the support for what you are asking, and a declaration that the desired result is the appropriate outcome.

The motion's conclusion is a succinct recitation of your request for relief but adds the conclusion that what you are asking for is the right outcome. Notice that the conclusion does not appear arrogant or off base in the context of what you have asked and the support you earlier outlined. It is a respectful request, but states why the outcome is both permissible under the law and proper given a summary of the facts upon which the prayer is requested.

CONCLUSION IN RULE 412 MOTIONS

In Rule 412 motions, I also suggest that you should copy the relief sought into the conclusion and provide its theory of relevance to the defense case. Some prefer to also include a rule citation, and to do so is not "wrong." I suggest at most to reference the subpart for which you are seeking admission.

For example, if the motion you are drafting seeks to admit evidence of a prior sexual relationship between the alleged victim and the accused, ensure a brief recitation of what evidence you want deemed admissible with a short reference (but not the rule itself) that should cause the judge to conclude the evidence is admissible. The conclusion of the motion need not reference the rule by number or by quotation. There will be plenty of other places in the motion to address the rule.

To illustrate, I offer, "For the foregoing reasons, defense moves this Honorable Court to admit evidence of the prior intimate relationship between PFC Smith and AV as evidence of consent or mistake of fact as to consent on the part of PFC Smith."

Often, motions that do not "win" in admitting Rule 412 evidence do so because they fail to "connect the dots" to explain how precisely the evidence is material to the defense theory of the case. The conclusion is one last place to use as an opportunity to be explicit in the materiality.

Another sample would be "The defense respectfully asks this Court to conclude that the prior intimate relationship between AV and PFC Smith constitutes admissible Rule 412 evidence as proof of PFC Smith's belief that AV consented on 31 October 2019.

Now that you have bookended your motion with what you are asking for (by the prayer for relief and the conclusion), you are ready to create the persuasive segment to convince the military judge to grant the favorable relief you are seeking.

Chapter 12

LEGAL RESEARCH IN RULE 412 MOTIONS

THERE'S ALWAYS MORE SO KEEP LOOKING

Never stop looking for legal authority.

There can be little doubt that judges are tired of reading the same verbatim caselaw in motions. Given that Article 120 cases have been the emphasis of the military justice sections globally since as early as 2007, there is an ever-expanding body of potential caselaw to mine for gold. So, please stop citing to the same cases over and over.

Do not mistake this as advice not to cite the seminal case or cases. *See* Ellerbrock[15] for example. Ellerbrock arguably is the most pivotal case for the defense in cases involving motivation to fabricate from fear of preserving a relationship and particularly a marriage. But there are cases since Ellerbrock that also weigh in on Rule 412 litigation.

I will continue to discourage anyone from starting "research" from the motions bank on the defense office's shared drive or even from those available from Trial Counsel Assistance Program (TCAP) or Defense Counsel Assistance Program (DCAP) as doing so can limit your consideration and cause a deficit in creativity. Relying on

---

[15] 70 M.J. 314 (C.A.A.F. 2011).

materials that have not necessarily been vetted can create additional problems of reliability. Have enough pride in your reputation and the client you represent not to rely on the prior work of others. As a military justice practitioner builds her own motions bank, extra effort put in early drafting motions the right way will pay off for future efficiency.

Judges read motions every week from counsel, and Rule 412 motions are frequently filed. They can tell if you have phoned it in and simply copied case law from motions in your office's brief bank. At minimum, Shepardize® the cases in motions *you* have previously filed.

In finding the authority for filing the motion already, the next source to perform legal research is to mine secondary materials such as David A. Schleuter's "Military Criminal Justice: Practice and Procedure"[16] or from a recent Criminal Law Desk book from The Judge Advocate General's Legal Center and School (TJAGLCS). Finally, you should move next to online search engines such as Lexis-Nexis® or Westlaw®.

---

[16] David A. Schleuter, Military Criminal Justice: Practice and Procedure (Matthew Bender & Co. 2019).

IDENTIFY SEARCH TERMS[17]

To identify search terms, always start with the Manual for Courts-Martial. If you have already followed my first book's suggestion (Shaping the Battlefield: How to Write Motions in Military Practice), you already have your first search term, which came from the Manual: the legal authority for the motion.

By identifying the legal authority from the exception to Rule 412 at the outset of drafting a motion you already facilitated the necessary legal research. The easiest mechanism to search for legal analysis on a given rule is to search for cases that include that particular rule.

Be mindful that amendments to Rule 412 and its subparts occur with a fair bit of regularity.[18] Know which subparts where your exception had been found and use the prior subpart for research. Merely moving its location in

---

[17] The purpose of this section in my book is not to provide a comprehensive class on how to conduct legal research in military cases, but instead to give some techniques to help get you started.

[18] In 2018, Mil. R. Evid. 412(b) was amended in part to more closely align with Federal Rule of Evidence 412. The amendment also addressed the Court of Appeals for the Armed Forces' opinion in United States v. Gaddis, 70 M.J. 248 (C.A.A.F. 2011) with regard to evidence the admission of which is required by the United States Constitution.

the Rule does not undo the body of case law on which that subpart was shaped.

The number of cases in a given search will be refined even by the addition of the Rule's subparagraph.

To ensure that your searches will only turn up cases that are relevant to military practice, either use a filter in your search engine or if you wish to look to federal circuits in addition to military courts, add the search term "and UCMJ" after each search parameter. Using "and UCMJ" may seem a bit elementary but many other jurisdictions use the acronym M.R.E. and only using M.R.E. will find cases that are not in the military setting.

To further refine the search, consider using "and motion" after the rule subparagraph.

To refine the search to cases that specifically will discuss whether or not the trial court abused his discretion, add the search term to ensure that the terms you are searching also include within the same paragraph the word "abuse" by typing "w/p "abuse". By adding the search term that it occurred within the same paragraph as "abuse" the number of cases shrinks from 677 to 207 cases.

To locate cases in which the reviewing appellate court analyzed whether or not the trial judge misapplied the law, add the search term (or substitute) "w/p error".

Chapter 13

PREPARE FOR LEGAL WRITING IN RULE 412 MOTIONS

DON'T MISS THE FORREST FOR THE TREES

Despite my insistence to do a deep dive in your legal research especially as it relates to more current cases, do not "miss the forest for the trees." Always cite to the controlling test.

In Rule 412 litigation, the constitutional exception rests on United States v. Gaddis, 70 M.J. 248 (C.A.A.F. 2011).

That said, do not stop with the controlling test. Do more.

HOW TO DO MORE WITH WHAT YOU FIND

When you obtain a workable number of cases, you will need to review them to ensure that the case stands for the proposition you wish it to. Careful review of the cases also ensures that the case is in the appropriate procedural posture to lend weight to your argument.

Recall that cases where a similar issue arose, but the appellate court did not analyze the issue because the court determined that the issue was waived on appeal because the trial defense counsel did not object are not particularly helpful. Cases where the issue was waived do little to support either side on a given issue because typically when an issue is waived, the appellate court will not provide any substantive analysis about the underlying issue. However, on occasion an appellate court will provide language in dicta that assists your cause by finding that error occurred but under the facts of the reported case, the error was harmless.

Any determination that error occurred is useful, but make sure that when you address the analysis on the case, you highlight the distinction from the facts of the reported case when compared with your own. The most significant distinction is often that the reported case prosecution had accused admissions or an outright confession so that any error did not prejudice the accused given the strength of his condemning statements.

In several military appellate cases involving Rule 412, the rulings announce that despite the appeal, that the trial court was justified in excluding whatever Rule 412 evidence the appellant wishes to appeal. Your analysis should not stop at this ruling. In many cases, the issue may have been waived on appeal if it was not raised at the trial court level.

In others, the appellate court may decide that the judge committed error but that given other evidence in the case it was harmless. You will see harmless error when there is other overwhelming evidence in the case, usually in the form of admissions or a confession.

If the type of evidence is similar to the evidence you want admitted in your case, these cases remain helpful to your own cause. If a case found error, you can use it to stand for the proposition that excluding the Rule 412 evidence was error. This is particularly true if the accused in your case did not make admissions or any statement to the authorities. Your research should not stop at the ruling. There is support in any case where error was found.

You can also find helpful and supportive case law in appellate opinions that stand for the proposition that evidence was properly excluded if you can distinguish the quality of evidence from your own. For example, if the defense in a given case had weak evidence of a relationship and the appellate court concluded the trial court properly excluded it, ask yourself whether evidence of a relationship in your case was much stronger.

Use cases where the appellate court said the trial court properly excluded evidence to distinguish your evidence from there. Case law provides ample fodder if you mine it enough. Do not stop at the holdings. Look for more.

CANDOR TO THE TRIBUNAL

Remember that because of the fundamental requirement of candor to the tribunal,[19] you have a duty to report case law that cuts against your position. If a case tends to support a different result than the one for which you seek judicial relief, comb through the facts and be prepared to distinguish the facts from your own case's factual circumstances in the analysis portion of your pleading. Do not ignore the case; do not rely on opposing counsel or the military judge to cite it for you. It exists, and you must address it.

---

[19] American Bar Association Model Rule 3.3.

Jocelyn C. Stewart

Chapter 14

LEGAL ARGUMENT (P-SAC)® IN RULE 412 MOTIONS

Some of the service courts separate out sections of their model-pleading format between the law and the argument, [20] whereas others combine them into one segment of a pleading.[21]

Realize that even in jurisdictions where the court rules require you to separate the law from your argument, you need to weave case law into your argument. If you fail to incorporate case law into your argument, the judge may not connect the dots that you want her to connect.

Fundamentally, the most effective motions are the ones where you are most clear on what you want, that what you are asking for is a legally available remedy, and then by supplying supporting case law that permits the judge to grant you your legally permissible requested remedy.

Law school writing instruction does not necessarily encourage the military writing style. Often military judges complain that the bulk of the motions they see are nearly entirely comprised of string citations of cases with little to no analysis or argument.

---

[20] Uniform Rules of Practice Before Air Force Courts-Martial, Rule 3.6(E) and Appendix B, September 2015.

[21] Rules of Practice Before Army Courts-Martial, Rule 3 and Appendix C, November 2013.

The argument format for military writing in motions practice that conveys the clearest message is one that follows the following "P-SAC" format:

- **P**remise
- **S**upport
- **A**nalysis
- **C**onclusion

Jocelyn C. Stewart

Chapter 15

PREMISE

PREMISE

What do you need the military judge to conclude in order to grant you the requested remedy?

The premise is a statement that sets the required standard in a persuasive, declarative statement. Without citing the rule or case that sets forth the standard, the premise statement weaves the standard into the statement with a tone that is persuasive but concise. The premise is based on the rule and / or a seminal case but should not be framed in terms of the rule or case.

Understand that there likely will be several premises in any given motion because a given form of appropriate relief typically has several factors that give way to the appropriate relief the motion seeks.

After identifying each premise, write out the declarative persuasive statement, one for each aspect of the legal standard. Write out each premise first before attempting to write out the rest of the section. By writing out each premise, you will automatically create the organization for your motion.

IDENTIFYING RULE 412 SUB-PREMISES

To identify the sub-premise, first identify the premise. Ask the question "What do you need the military judge to conclude in order to grant you the requested remedy?".

Within each major premise, there likely are to be several sub-premises. Once you identify the factors or standards for a given motion, each of those factors or standards likely forms a sub-premise. The sub-premise is the factor or element of the informing legal standard. To identify the sub-premise for a given major premise, you ask yourself, what are the factors or conditions that the Military Judge needs to conclude in order to conclude the major premise.

In Rule 412 motions, the major premise is that without evidence of either the alleging victim's sexual history or predisposition, the military client will not have a fair trial.

In Rule 412 motions, the sub-premises are each of the components required for introduction. This includes a balancing test.

Realize that in light of <u>Gaddis</u>,[22] there is a separate test (and therefore different sub-premises) when the evidence a party seeks to introduce is being offered under the constitutional exception. Realize that <u>Gaddis</u> and the 2018 amendments directly reject <u>United States v. Banker</u> so be mindful if you find caselaw that relies on <u>Banker</u>.[23]

The sub-premises generally for a Rule 412 motion are:

1. The evidence (being offered) is material.
2. The evidence (being offered) falls within an exception to the general prohibition to M.R.E. 412 evidence.
3. The probative value of the evidence (being offered) outweighs the danger of unfair prejudice [to the alleged victim's privacy].[24]

---

[22] 70 M.J. 248 (C.A.A.F. 2011).

[23] 60 M.J. 215, 223 (C.A.A.F. 2004).

[24] *See* Manual for Courts-Martial (2019) Analysis to the Military Rules of Evidence, Rule 412.

Chapter 16

SUPPORT

SUPPORT

What are the rules and cases that support your premise?

Support is not merely your legal conclusion that it is admissible. Support means what forms the basis to permit the judge to do what it is you are asking her to do.

String citations without analysis are not particularly persuasive. Avoid stringing citations together for an entire paragraph and especially do not string together citations for multiple sequential paragraphs.

Even though support is a crucial component of any legal pleading, remember not to inundate or saturate the Military Judge with string citations. Any reader loses engagement without the author providing her with some type of explanation why the given case or other form of support assists with the major premise or sub-premise. Even in complex motions, a drafter should cite to one rule, perform analysis and application, before turning to a different rule or source authority.

That explanation or orientation to the facts of a particular case is what is encompassed in the analysis and will be explained in the next section.

When you are drafting a motion involving a particular subpart to Rule 412, you should separate out each grouping of cases or rules into subsections, which are those sub-premises just discussed in the previous section.

Your primary source for support in P-SAC® is Rule 412 and its component parts. Citing to each exception to Rule 412 for which you are seeking admission will assist the judge in understanding materiality and in orienting the judge to her ability to grant your requested relief.

If possible, also consider using the seminal case as support because there is always caselaw that has analyzed a Military Judge's previous application of certain facts to that particular subpart or exception.

For example, if you are drafting a motion under the constitutional exception to Rule 412, cite to <u>Gaddis</u>. Do not cite to <u>Banker</u>. I guarantee you that 95% of the motions from your office motions bank cites to <u>Banker</u> which was specifically repudiated in <u>Gaddis</u>.[25]

---

[25] 70 M.J. at 254.

Jocelyn C. Stewart

Chapter 17

ANALYSIS

ANALYSIS

How do the facts of your case fit the support you provide to lead the reader to conclude that your premise is valid?

Analysis is often the most neglected area of anu military motion and particularly is absent in Rule 412 motions. Analysis is the part of a motion that cannot be copied and pasted from another person's pleading. When copying and pasting, parties generally cut out the analysis but never insert their own. This is a treacherous mistake especially in Rule 412 motions.

Recall that Rule 412 is a rule of exclusion. The Military Judge will not infer anything in a Rule 412 motion, so you have to be particularly painstaking in your analysis. Take it step by step, be elemental. You may feel silly, but you are setting yourself up for a significant increase in your chances of succeeding in your motion.

Analysis tells the Military Judge why the result in a given legal case (admission of Rule 412 evidence) applies to the facts of your case. If the judge excludes this evidence tell him in the analysis why this case likely would be overturned. Judges do not enjoy being overturned, so spell it out for her.

The analysis section of your motion communicates to the Military Judge that not only have you well-researched the area of law, but also that you have read the law, understand the law, and are applying it to the facts of your particular case.

I have read dozens of motions and even more motion responses that indicated to me its author used a series of string citations without ever reading one of the cases cited. When I spend the time reading the motion or a motion response so that I can prepare for oral argument,[26] more often than not the facts of the case are distinguishable from the facts of the case being litigated. In some instances, the case actually supports my position, and not opposing counsel's. It seems absurd to write this in a book, but it bears stating. Unless you have read a case, you should not cite to that case.

I understand better than most the demands of one's time as a litigating trial lawyer, balancing the requirements of responsiveness to clients and family. There is time to

---

[26] *See also* Shaping the Battlefield: How to Litigate Motions in Military Practice.

read each case; it may take you longer the first few times you draft a motion, but there will be repeat issues and your learning will amplify when you make the time to read the law.

Analysis is more than string citations and then declaring at the end that your premise is supported. Analysis tells the Military Judge (and your opposing counsel) why your remedy is appropriate under the facts of your case.

An example of the analysis portion of a Rule 412 motion:

Evidence that AV was self-employed on OnlyFans® and that ACCUSED made her terminate the employment is material because it provides AV with substantial motivation to fabricate her allegation against ACCUSED. Evidence of bias and motivation to fabricate falls squarely within the constitutional exception to Rule 412's general prohibition against evidence of sexual behavior and predisposition of the alleging victim. Evidence that AV was self-employed on OnlyFans® and the ACCUSED made her terminate her lucrative employment is akin to that which was deemed constitutionally required in United States v. Ellerbrock, 70 M.J. 314

(C.A.A.F. 2011) because this evidence represents bias and motive to fabricate. The same is true for ACCUSED finding AV's lingerie and sexual toys in her room during the barracks instruction and shaming her for it, especially in their aggregate, and given the timeline of events leading up to AV's allegation against ACCUSED. In the backdrop of AV's dissatisfaction at living and working on INSTALLTION, she becomes pregnant and marries her child's father; still, she is prevented by the Army from moving and having the life she wants. At the helm of her associated problems within the Army is ACCUSED (SUPERVISOR), yells at her and shames her for trying to secure additional income and then yells at her for sexual items in her barracks room which she largely believes are not his for comment. Finally, under the defense theory, she has had enough with being shamed and yelled at by ACCUSED, and she reports him for a fabricated assault. While potentially embarrassing to AV, this evidence would satisfy the required balancing test even were it not constitutionally required because AV publicized not only her employment on OnlyFans® and was openly hostile about her being made to end it by ACCUSED. AV made her employment on OnlyFans® known to her peers and even ACCUSED as her supervisor. The materiality and probative value are exceptionally high in explaining why AV was so enraged with the Accused as to knowingly file a false claim of sexual assault against him.

Under the framework of <u>Ellerbrock</u> and M.R.E. 608(c), each piece of evidence is not only logically and legally relevant, but also is material, falls within the constitutionally required exception to Rule 412. Without this evidence, ACCUSED cannot have a fair trial.

The analysis portion of your motion necessarily weaves the legal standard into the facts of the case. In the above example of a motion to admit Rule 412 evidence under the constitutional exception the author of the motion lays out what is the legal standard. The analysis describes exactly why the evidence is material to the defense case, pinpoints the rule's subpart to permit its inclusion, and identifies why the balancing test is satisfied.

Many Rule 412 motions state the conclusion that the balancing test is met without stating why or how. Be specific, even if you feel like it should be obvious.

After describing the legal standard with support from case law and / or the code, analysis begins. The analysis lays out the facts that support inclusion. Analysis lays out what the facts are (that you stated initially in your facts section) to support that the evidence is material, meets an exception, and satisfies the balancing test.

Now that the facts of your particular case have been applied to the legal standard, in the last framing of the analysis, you should loop back in the case law. Here, we remind the Military Judge that case law gives us a definition of what is an official inquiry or questioning. Then we make a declarative and conclusory statement why our facts applied against the law leads to the answer that our motion is arguing: there was a violation, and the result is that the statement is inadmissible.

Do not mistake the motion's conclusion for the declarative, conclusory statements that are made at each segment of analysis. In some instances, there may be a summary repeat of those declarative statements in the sections of the analysis portion of the motion, but they are not identical. The analysis from the motion builds to the conclusion of the motion that the law supports the application to the facts that you are arguing.

Often and in nearly all motions there should be more than one section of analysis. In the above example, a motion writer could easily break the predicate facts and law into two separate analytical paragraphs. The first predicate to our desired judicial conclusion that the Rule 412 evidence should be admissible is its materiality. The second predicate is that it meets a specific exception to the Rule. The third predicate is that it satisfies the balancing test.

Too often military lawyers who engage in motions writing view the above type of analysis as too simple or overly juvenile. I promise that most military motions do not apply analysis, and I suspect that their authors assume the judge will apply the facts of a given case, as they want them to without needing to help them. This represents a wasted opportunity. Help the judge help you!

Do not assume the judge will reach the same conclusion unless you help "walk the dog." Write in a way that leads any reader to reach the conclusion you want them to by supplying them with the law but also applying your facts to the law. String citations do not help anyone unless you have explained why those cases support each analysis' section legal conclusion you want in the context of the facts of your case.

Writing in a systematic way does not make you look like a simpleton; it makes your conclusions and therefore your ultimate prayer for relief seem like the easy conclusion. I promise that advocating your position in a systematic way that includes analysis and does not assume it has transformative potential.

Jocelyn C. Stewart

Chapter 18

CONCLUSION

CONCLUSION

Applying the FACTS of your case to the supporting rules and cases, your premise is the right result.

The conclusion, in this sense, is a declarative statement at the end of each analysis paragraph that is the logical conclusion to that analysis. For example, in looking to the analysis examples in the preceding segment of this chapter, the conclusion is as follows:

Under the framework of Ellerbrock and M.R.E. 608(c), each piece of evidence is not only logically and legally relevant, but also is material, and falls within the constitutionally required exception to Rule 412.

I included the conclusion for the analysis in the preceding sample of analysis to indicate where to place the conclusory statement. Please note that given what precedes it, the conclusory statement may seem obvious. It should be because the example "walks the dog." Its clarity is not a reason to omit the conclusion.

Many legal writers assume the conclusion of each analysis section. The warnings of Rule 412 motion writing should guard against shortcuts. Remember this is a rule of exclusion, and the judge is looking for a way to exclude what you are proposing to include.

Jocelyn C. Stewart

Chapter 19

ORGANIZATION IN RULE 412 MOTIONS

No matter how well reasoned your legal argument / analysis section, your arguments are bound to be lost in a sea of paragraphs unless they are organized around headers. A header is a demarcation of what topic you are discussing. I recommend making each header a declaration of each legal argument you are crafting. This process means you are organized and providing a persuasive framework.

Choose your headers and the framework of your argument based on the standard set out in the law for the particular area of your motion. For instance, if the motion you are writing is about a prior sexual relationship between the putative victim and the military accused, the headers should mimic the sub-premises. Use declarative brief sentences as the headers. Underline the headers so they are more readily identifiable.

<u>The Evidence is Material</u>

<u>The Evidence Qualifies for Admission under Rule 412(b)(2) to Prove Consent or Mistake of Fact</u>

<u>The Probative Value of the Evidence Outweighs the Danger of Unfair Prejudice to AV's Privacy</u>

Each of these suggested headers is a sub-premise in the analysis that helps your Military Judge to know whether the evidence should be admitted. Each section has a body of case law that helps us to know whether a given scenario has a certain legal consequence. Your analysis should carve out a separate section, which is annotated by a declarative header.

Jocelyn C. Stewart

Chapter 20

WRAP UP IN WRITING

As in most aspects of litigation, motion writing has elements of art and science. The systematic approach in this book for drafting a Rule 412 motion in military practice provides the organization and framework to phrase your requests for relief in a way most likely to achieve each desired end. I would argue this book's aim is to aid you with the science. The art lies in the tactical decisions that led you to your determination that you should write the motion.

I'll add that drafting (much of) a Rule 412 motion has on more than one occasion led me to shift my tactics and to really hone in on what facts qualify as Rule 412, and which do not. I am always strategizing about which facts I need not include for the best tactical advantage, but never will I omit facts that must be pleaded under Rule 412. Playing "fast and loose" with Rule 412 is unethical and dangerous to the client.

Motion writing is not drafting opening statements. Motion writing is not creation of closing arguments.

Motion writing is not academic writing. Let me repeat for emphasis. Motion writing is not academic writing. Use of verbose or flowery language will not draw favor or accolades.

In addition to the foundation of organization I will also advocate that each of you builds in time to your process to draft and then to step away from the product before editing. I would prefer this time be broken up by sleep, but if not possible, at least interrupt the process with another unrelated task or better an activity away from your desk. Your writing and reason will be better for it.

I am excited to finish this next phase of my project. My goals will always be the betterment of military court-martial practice. Your client, whether the command or the servicemember client, deserve the very best you can give them. I trust this book will serve that pursuit of delivering your best.

Jocelyn C. Stewart

PART II:

HOW TO LITIGATE

RULE 412 MOTIONS

Chapter 21

YES, YOU NEED AN ARTICLE 39A SESSION

THE RULE SAYS YOU MUST

The Court cannot rule on 412 evidence without conducting a closed session. Rule 412(c)(2) states in pertinent part that "Before admitting evidence under this rule, the Military Judge _must_ conduct a hearing, which shall be closed" (emphasis added).

This means that every proponent of a Rule 412 motion or response to a motion seeking admission of Rule 412 evidence must request a closed Article 39a session of the Court.

WHY THE GOVERNMENT NEEDS AN ARTICLE 39a

If you work for the government, you must give proper notice to the alleging victim and her counsel, if she has one. You also must provide an opportunity to be heard by the alleging victim directly or through her counsel. You will need to capture on the record that the putative victim was notified both of the existence of the hearing, its date, time, and place, and that you notified her of her opportunity to be heard. The judge will also ask whether she desires to be present or to be heard.

WHY THE DEFENSE WANTS ONE ANYWAY

If you are the defense, you will in all likelihood not have access to the complaining witness in advance of motions litigation or even in advance of trial.[27] In light of this statutory reality, because Rule 412 motion litigation necessarily involves the sexual behavior and / or sexual disposition of the alleging victim, you will want an opportunity to examine the putative victim.

If the putative victim denies the underlying conduct or predisposition under oath, realize that you will need to modify the basis for which you are seeking introduction of the Rule 412 evidence: to impeach the putative victim.

OVERALL LANDSCAPE CONSIDERATIONS

Do not use oral argument because you like the sound of your own voice. Very likely your judge does not, and it doubtful that opposing counsel does. If you have followed the instruction and methodology from this book and its predecessors, the judge will have most of what she

---

[27] Article 6b, UCMJ, codified elements of the Federal Crime Victim Rights Act. Subsection (f)(2) of Article 6b outlines that when counsel for the accused requests to interview an alleged victim of an offense, the interview "shall take place only in the presence of the counsel for the Government, a counsel for the victim, or, if applicable, a victim advocate." 10 United States Code Section 806b. It has become fairly common place for alleged victims to decline any pretrial interview.

needs to render her decision from the pleadings. The only missing piece may be from the putative victim and or the defendant client.

Do not request a hearing and then not offer any additional evidence or fail to provide any oral argument. No one wants to feel that their time is not valued or respected. Even though a Rule 412 motion hearing is required, view each session as an opportunity to ensure the judge has all of the evidence that he needs to fairly decide the motion.

Chapter 22

RIGHT TO A PUBLIC TRIAL

BE CONTIENTIOUS OF YOUR CLIENT'S RIGHT TO A PUBLIC TRIAL

Rule 412 motion litigation will require a closed session.

However, be cognizant that your client has a right to a public trial. When motions overlap and some facts apply to both Rule 412 litigation and other factual motions, be cautious of ensuring your client's fundamental right to a public (and therefore fair) trial remains sacrosanct. The founding fathers of our nation realized that process occurring in secrecy is a close cousin to injustice. Transparency breeds fairness, and because military courts-martial practice is "command" driven, one can argue its public nature is even more important to maintain societal confidence.

As you guard against secrecy and injustice, you must ensure that closed sessions are only utilized for matters that require them to be so. For example, when multiple motions are being litigated, counsel and even judges can create missteps that preclude your client from receiving a full public trial.

One common mistake is to be tempted to make oral argument about non Rule 412 matters in a closed session. Realize that oral argument during those closed sessions should only be made on matters that are appropriate for consideration in those sessions, namely, on Rule 412 facts.

What happens in some cases is that there is overlap between facts elicited during the closed session that bear upon a motion that would not require a closed session. Do not be drawn into making oral argument on matters that do not pertain to Rule 412 in a closed session.

If you realize that the Court is closed but should not be, do not be afraid to respectfully address the issue. The judge will be thankful that you did.

Jocelyn C. Stewart

Chapter 23

PREPARING TO LITIGATE RULE 412 MOTIONS

PREPARATION IS KEY

Anyone who wants to win a motion or at least best position their chance for success needs to prepare to litigate that motion.

Preparation includes refamiliarizing yourself with the law that supports your position. Preparation means knowing the legal standard required to be victorious. And preparation necessitates you know the facts that matter to the law on the briefed issue.

REREAD YOUR OWN MOTION

In most instances, between the time you wrote your pleading and the time to litigate the motion you may have handled several different issues on several different types of cases. Go back and reread your pleading.

Additionally, reread the seminal cases that matter to the requested prayer for relief. Identify if you missed anything that you should have included. Determine whether a new case that came out since you wrote the pleading.

Reconsider if there are facts you need to supplement in order to best position yourself for success.

READ AND DISSECT THE RESPONSE BRIEF

Perhaps the greatest mistake a counsel can make is to not read the response brief, or to read it but not dissect it.

First, reading opposing counsel's brief will help identify the areas of disagreement to refine the litigation. Realize *before* the night before your Article 39(a) that if the opposing party conducted further interviews or investigation, there may be additional facts you will need to contradict during the hearing. Do your homework and know what is coming. Be prepared to elicit even more evidence or additional clarifying information from an affiant or fact witness.

Second, reading opposing counsel's brief will direct and guide the additional legal research that you will need to conduct in preparing for oral argument. This is the part of the preparation for dissection of the reply brief. Do not be so arrogant as to believe your own legal research could not have missed something critical. Do not assume that you found every case or that your opponent has not identified a case from federal case law that you did not do a deep dive to discover. But also, do not assume that when the government cites a certain case for a particular position, that they actually read the case, or that the case stands for the stated proposition.

Chances are, that government counsel did not read my first book: ***Shaping the Battlefield: How to Write Motions In Military Practice*®** to heed my warning against drafting motions from a brief bank. If the government counsel did not read or apply the principles in that book, then chances are they have not read the cases they referenced in their motion, and instead they "cut and pasted" string cites. There often is gold in those string cites for their opposing counsel's position. So be sure to mine it.

PREPARE TO MEET THE RESPONSE EVIDENCE

After you read and consider the response pleading, you will need to understand what evidence the government intends to elicit during the hearing, what evidence they proffered, and which pieces of evidence they appended to their motion.

Review any listed witnesses in a reply brief. If the government listed one or more witnesses, your preparation must include interviewing each of their intended witnesses. Reach out to the witnesses sufficiently in advance of the motions hearing so that if (and when) the witness elects not to respond to your inquiries you can reach out to opposing counsel to ask that the counsel facilitates your access. If the witness refuses, be prepared to make your record. In many instances, the judge will not require the witness to submit to a pre-hearing interview. However, I have found if you let the judge know on the record prior to calling the witness to the stand, that he will be more amenable in granting fairly wide latitude during examination.

Analyze the government reply brief for proffers of evidence. Often a government counsel's writing in their reply brief will give you a strong indication whether the putative victim has denied the behavior your motion is attempting to make admissible at trial. I classify this as what the motion response proffers as evidence. In all likelihood, the government counsel interviewed the alleged victim about the subjects of your motion. Make sure you get a copy of any discovery about that interview. But also annotate in the motion response what the counsel is intimating as will be the alleging victim's version of events for the motion. I have seen on numerous occasions that what she says on the stand is not consistent with what was proffered in the government brief. This is another reason obtaining discovery about the government interview is critical for later potential impeachment evidence.

Lastly, review all appended evidence that the government lists in their response pleadings. Sometimes, the counsel only repeats a piece of evidence you already included in your motion. Others, the government counsel conducted additional interviews or investigation in response to your motion, and this evidence could reveal a significant undermining of your recitation of the facts. Never take this evidence at face value; this is not to say the government is providing false information to the Court. What I mean is that there could be another set of witnesses who saw the events differently from the one the government provided. Investigation does not stop at the filing of your motion. Review, analyze, and if needed, get back out there to keep digging for more.

GET ORGANIZED

Depending on your judge, you may encounter a "hot bench." What I mean by a "hot bench" is that the judge will pepper you with questions before you ever have a chance to make a scripted prepared speech for your argument. Often Military Judges have already honed in on the areas they feel they must resolve to make their ruling. Those areas for further review will guide the Court in their dissection of your motion. That often results in counsel being peppered with questions, rather than being given an opportunity for planned soliloquies.

To best position yourself to be able to handle those questions, you will need to be organized. I will typically have saved the cases I want to be prepared to provide insight regarding pulled up in multiple windows on my laptop. I may have the pleadings themselves printed with highlights or notes. Whatever your method, the key will be to know where the cases, pleadings, and evidence are to be able to use them if you need them. Being organized is more than half of your preparation, especially if you will encounter a "hot bench."

Jocelyn C. Stewart

Chapter 24

EVIDENTIARY TACTICAL CONSIDERATIONS IN RULE 412 MOTIONS

EVERY MOTION REQUIRES EVIDENCE

Recall that the recitation of facts in your motion is not proof of the underlying fact. You need evidence. A proffer of counsel is not evidence.

NOT EVERY MOTION LITIGATION REQUIRES ADDITIONAL EVIDENCE

If you drafted the motion as you should, every paragraph in your facts section is verifiable from a written document you reference in your motion as a marked appellate exhibit or by written affidavit provided to the Court at the time of filing (also an appellate exhibit) or provided to the Court at the time of litigation on the motion. *The only additional evidence will come from calling the putative victim.*

As you are deciding what additional evidence to provide, you will need to consider any holes in your motion, any need to respond to evidence the government supplied in its reply pleading, or to confront witnesses the government intends to call during motions litigation.

ASSESSING THE BEST AVENUE FOR YOUR TACTICAL CONSIDERATIONS

Given the options that you have in documents, affidavits, scholarly work, and potential testimony, you will need to assess and anticipate the benefits and drawbacks of each.

Realize that if you provide a signed, sworn statement by a witness, future impeachment is easier from a logistical standpoint. Providing that signed, sworn statement with the pleading means that opposing counsel has additional time to respond to it. On the other hand, the signed, sworn statement by a witness gives your judge clarity on the facts and gives the judge more time to connect the dots in your motion. The affidavit also pins down the witness about their recollection before they can decide they do not want to "be involved" or change their minds after the passage of time or even after speaking to the alleging victim.

Never offer evidence unless you absolutely need to do so. By this I mean, if you conduct a pre-hearing interview of a witness, that witness may have knowledge about certain facts which are germane only to the motion and maybe the same witness knows additional facts that are only relevant to the trial. There is no requirement to reduce to writing all of the information that the witness possesses. All that is necessary to provide in the motion and the potential affidavit are the facts that are relevant to the motion. With time, these considerations may become almost second nature.

I encourage you to make these decisions as part of a team. There may be considerations you had not contemplated that your co-counsel sees and vice versa. Never allow experience to cloud you from being open to the idea that less experienced counsel may see issues you had not considered.

Chapter 25

DOCUMENTARY EVIDENCE IN RULE 412 MOTIONS

DOCUMENTARY EVIDENCE: YOU HAVE OPTIONS

One of the first questions a Military Judge will ask during an Article 39(a) session for litigating the motion is whether or not the parties object to the Court's consideration of the evidence already offered at the filing of the motion.

The next question the judge typically asks will be whether the litigants have any additional *documentary* evidence to present on the motion.

Among the options that you have is to furnish any additional reports or segments from the investigation that you may have inadvertently omitted in your initial pleading. Perhaps the government provided you with additional discovery after the time of your filing deadline. If at all possible, please send these supplemental documents to the court reporter for marking at least the night before the hearing or in the morning prior to going on the record.[28]

---

[28] Know and adhere to any local rules or other limitations outlined in the case pretrial order on when you may request that documents be marked.

USE OF AFFIDAVITS

Another option is to provide the Court with an affidavit. An affidavit is a written declaration with some form of attestation or swearing. Affidavits are permitted pursuant to Rule for Court-Martial 905(h).

From my experience, submission of sworn affidavits underutilized in motions practice (and in court-martial practice generally). Affidavits are a practical, useful instrument to control the information that comes to the Court for its consideration.

A subset of underuse generally by either party is the underuse of an affidavit from the defense client, the accused. There are substantial reasons to submit evidence from the defense client in this methodical presentation, rather than calling him to the stand and risk missteps on cross-examination. In Chapter 28, I will discuss the tactical considerations of whether to call the client to the stand, which I believe are more often far outweighed by the risk in calling him to testify during motions' practice. Counsel forget that an affidavit provides oft-needed evidence without risking the issues pertaining to the motion or to trial strategy.

When using affidavits from witnesses other than the defense client, I typically file those with the motion. If a filing suspense is looming and the witness has not returned the final signed affidavit, but the witness verified its accuracy over email, I will file an affidavit by the paralegal working to obtain the signature. If instead you request a placeholder for the anticipated future signed affidavit, Courts will often indicate they will not consider the evidence on the affidavit until signed. Either way, I believe it is still best practice to give notice that the evidence is anticipated.

In Rule 412 motions especially, I believe it is the best tactic to provide evidentiary affidavits from non-party witnesses with the motion itself. The Military Judge has to make a preliminary determination to hold the closed session. If the motion that is filed does not have any Rule 412 evidence in support of the motion, the judge could decline to even hold the closed session.

Recall that if you decide to use a client affidavit, you are not required to file it or even reference it with your motion. The same is true whether you intend to call the client during the closed hearing or not. Given that you are not required to file notice of your potential intent to call the client at the motion's hearing, you can feel comfortable providing the client's affidavit when the judge asks for any additional documentary evidence.

The Military Judge will be far more patient with you if you provide an affidavit from the client to the judge on the record if you gave it to the trial counsel at least a few minutes before going on the record. Because you are not required to inform the judge or the government prosecutor when you intend to call your client to give sworn testimony, providing an affidavit later in litigation is hardly worth complaining. If the government counsel complains, and if the judge asks you for your position, do not oppose a brief recess or even a recess in place for him to review it before proceeding.

Jocelyn C. Stewart

Chapter 26

WITNESS EVIDENCE IN RULE 412 MOTIONS

WITNESS TESTIMONY

There are reasons to call no witnesses on one particular Rule 412 motion, and there are reasons to call several witnesses on another. As with any litigation decision, be intentional in making your choice, and realize there is no set rule. Every Rule 412 motion litigation is different.

Similarly, there are significant decisions to be made in which witnesses to call, if any, and the preparations to be made prior to calling any or all of those witnesses.

FRIENDLY WITNESSES

When witnesses are "friendly" to the defense and I have made a tactical decision to call that witness to testify, I will always prepare the witness in advance of the hearing. That preparation includes the information I intend to elicit, though not my precise questions. Recall your discovery responsibilities if you provide any materials to a witness to help them prepare or to refresh the witness' recollection.[29] In practice, it is seldom helpful to your position to provide your notes or questions to a witness as part of preparation given the duty to then provide them to your opponent.

---

[29] Rule for Court-Martial 612 outlines obligations for "an adverse party" which includes government and defense counsel.

NON-FRIENDLY WITNESSES

When witnesses are not "friendly" I do not tend to prepare the witness in the same way that I would a "friendly" witness. In some instances, I will not speak to the witness at all. On some occasions, that is by choice, whereas for others, it is not.

Recall that there are also occasions when I am powerless to speak with non-friendly witnesses in advance. The most notable of these is the complainant. There may also be relevant evidence from the complaining witness' family member or close friend. Please see the discussion that follows in Chapter 27 of this book that deals specifically with tactical considerations on calling the complaining witness in a case for purposes of litigating Rule 412 motions.

WITNESSES GENERALLY

As in trial, because they are human, witnesses are not always predictable. Despite preparation, a witness may not provide evidence that you anticipate they would. Note anticipation is not guesswork, but it is built on preparation and to the maximum extent possible from prior interview and from reviewing prior statements made to law enforcement or from government disclosed notes of their pretrial interviews.

In addition to the human element of calling a witness, there are other tactical considerations such as whether opposing counsel will have an opportunity to cross-examine your witness prior to trial. Note these are some of the parallel considerations about why you may decide against wanting to call your own client on a given motion. More on that topic will follow in Chapter 28.

Again, every case is unique. Ultimately, the individual facts of your case will be your guide in making these critical decisions.

Chapter 27

CALLING THE COMPLAINANT AS A WITNESS IN RULE 412 MOTIONS

THE COMPLAINANT'S RIGHTS

Understand that the statutory right that a complainant has not to be interviewed in advance of trial does not extend to interlocutory matters for consideration on the record. A complainant cannot refuse to testify at a motions hearing where there are matters in dispute. Please note that in almost all situations, there are factual matters in dispute from the complainant's and defense's standpoint, especially with regard to Military Rule of Evidence (M.R.E.) 412 matters.

YOUR ONLY OPPORTUNITY FOR QUESTIONING

So long as you have a good faith reason to call the complainant in a motion, I would encourage any defense counsel to do so. Under the current rules, this may be the only opportunity a defense counsel has to pose any questions to a complaining witness prior to the complainant taking the stand at trial. Because you are calling the complainant as your witness in a motion where you bear the burden of persuasion, your questions should begin in a direct examination format.

Many Military Judges encourage more pointed, directive questioning during Article 39a sessions. In Rule 412 motion litigation, I encourage practitioners to ask open-ended questions because it often means the complaining witness will talk more than she would otherwise. The more, the better when it comes to pretrial litigation.

In most instances, the complaining witness will not be well prepared to answer open-ended questions, and you will benefit from allowing the witness to provide as much information as the complainant is willing to provide. Be patient. Initially their answers may be short, but with deliberative questions, intentional pauses, and open body language, I have found a complainant will eventually reveal more.

TONE AND TENOR

During litigation of Rule 412 motions, consider that the way you interact with the complainant will set the tone for additional potential interaction. There are reasons to use more pointed questions during other motions litigation. This may inspire a harsher tone on the part of the complainant at trial that may seem incongruous to the panel at trial, which may give you an advantage in trial. The panel's only experience is in front of the team where you are being gentle but inquisitive.

You should consider whether your tone and tenor will inspire the complaining witness to agree to a future interview prior to trial. There likely are many reasons not to ask pointed questions during motion litigation. Every case is unique, and individual strategy will dictate the best strategy to employ. As I always encourage, be deliberate in whichever approach you take. Litigation is not about accidents; it is about thoughtful and careful consideration and choice. Deliberation and intention assist practitioners in shaping their battlefield.

USE THE GIFTS YOU ARE GIVEN

In nearly every case the government counsel will have had access to the complainant witness at least through the alleged victim's attorney. From your review of the government's response pleading, you will have a pretty good appreciation of which facts you allege in the Rule 412 motion that the complainant disagrees with. Even if you believe the complainant will testify under oath to contradict your motion, this helps you in your motion.

Let me say that again. If the complainant denies the underlying conduct you allege in the Rule 412 motion, this provides a separate basis for its introduction. The evidence you seek to elicit now becomes relevant to impeach the complainant by contradiction.

To help concretize what I mean, I will provide an example. In a contested Article 120 case alleging sexual assault and a forced kiss (yes, you read that correctly), there were at least three witnesses who had provided sworn affidavits to me prior to motions litigation that each had observed grinding type dancing between my client and the complainant on the same night and in the same timeframe as the charged offenses. On the stand during motions litigation, the complainant testified under oath that their bodies never made physical contact other than her hands clasped around the back of my client's neck while dancing. About five questions later, the complainant realized she had told the police that my client had kissed her while they danced, which she claimed she had rebuked. She changed her story on the stand, and of course it was under oath.

BE READY TO USE DENIALS AS ADDITIONAL BASES FOR ADMISSION

During oral argument, I informed the Court based on the complainant's denial of the contact under oath and the eye-witness evidence that contradicted her account, defense was now offering the evidence under an additional basis. We were already offering the evidence as a matter that bore upon the client's state of mind that their mutual flirtatious and sexual dancing made him believe she was receptive to sexual contact. Defense was also offering the evidence as part of the complainant's motive to fabricate, that fellow Soldiers had taken notice of the inappropriate contact between her and her squad leader (the accused) and had warned her that she should knock it off before she got into trouble for fraternizing. She had been drinking alcohol underage, and defense posited when she realized the next day the risk of repercussions for her behavior, it prompted her to report their contact as non-consensual.

Now, because the complainant also denied the grinding, we offered the evidence to impeach her by contradiction. The Court ruled that the evidence was admissible for all proffered bases. The judge ruled from the bench. Note this case never went to trial.

After leveraging every favorable motion and circumstance in the case, we reached alternate disposition with no conviction. We shaped our battlefield and avoided the war altogether.

LATER USING THE COMPLAINANT'S TESTIMONY

Remember that trial litigation is chess, not checkers. Rule 412 motions practice is no different.

After the complainant testifies, ensure that you obtain a usable copy of the testimony. If the complainant testified during a closed session, whether pursuant to Rule 412 or possibly Military Rule of Evidence 513, or both, ensure that the Court grants a separate but related motion to provide that material to the defense as evidence necessary to the preparation of the defense.

The best practice is to ask the Court to unseal the audio-recording and separately to allow the defense to supply to a private court reporter for transcription. In several cases, the judges have denied release to anyone outside the defense team and even to permit the defense to type its own transcript. Know your judge, but always ask for what you need to be able to effectively impeach the witness.

Impeaching the complaining witness with prior sworn testimony can be pivotal in the case, and you will not want to stumble into an objection that the material is not permitted because it arose during a closed session of the Court. Game time calls in trials involving previously closed session will seldom go to the defense.

In a trial at Fort Wainwright, Alaska, a Military Judge shut me down on asking questions that related to information elicited during Rule 412 litigation that was not

Rule 412 evidence. As a result of that ruling, I build into my Rule 412 litigation a request to use evidence which is not Rule 412 evidence from the Article 39(a). After employing this practice, I have not had a Military Judge deny me again.

The sooner you deal with these issues, the better. Give your judge time to consider your position. In a case I tried before a Military Judge who was sitting on his first contested panel case, he delayed ruling on the matter until the first day of trial and denied our request to release the matter to a privately hired court reporter for transcribing. Eventually he released the audio, but we were not permitted to make even our own verbatim notes. It made matters more difficult (though not impossible) for impeachment, and I regretted not raising the issue sooner.

Even if not granted the ability to send the recording to a court-reporter, by asking early enough, you will be able to formulate notes or nearly verbatim notes with time hacks for impeachment at trial.[30]

---

[30] In the future, look for additional series in trial litigation techniques.

Chapter 28

CALLING THE ACCUSED AS A WITNESS IN RULE 412 LITIGATION

GOVERNMENT COUNSEL CANNOT DO IT

To be very clear at the outset, only the defense may call a military accused during a motions hearing (or military trial). It may seem silly to clarify this point, but I promise I have seen a government counsel attempt to call a military accused to testify on a motion in two separate cases. Don't be that counsel.

DECIDING WHETHER TO CALL YOUR CLIENT

Often the military accused is the only witness to one or more facts necessary to litigate a Rule 412 motion. The defense client, indeed, may be the only person willing to attest to prior sexual activity between him and the complainant. Perhaps the accused person is the only person aware of an intimate detail that the complainant once confided in him about a prior trauma or sexual assault.

Defense attorneys will often need to establish the impact that knowledge of those facts had on the military accused person's belief about consent. The factual predicate includes the client's knowledge of the complainant's sexual behavior *before* their encounter. The factual predicate also includes confirmation that this prior knowledge impacted his belief about consent. Otherwise, you have not provided the requisite factual predicate as to why (the materiality) a given prior sexual activity is necessary to your client's Constitutional right to confront his accuser.

AVOID SWORN TESTIMONY AT ALL COST

Since 2004, I have seen defense counsel come to believe that government counsel will not be prepared to cross the military accused, and therefore a defense counsel should risk calling the accused to the stand "for the limited purposes of the motion."

Please note that the words "for the limited purpose of the motion" are not magical words. Those words have no bearing on whether or not your client can be caught off guard by questions posed to him. Those words do not guarantee that you and your judge will have the same impression of what aspects of potential testimony go to the motion and those that regard the alleged misconduct. That phrase does little to protect a client regarding perceived inconsistencies in his future potential testimony at trial. Also, gravely, a client's answers can paint your defense into a proverbial corner from a strategic standpoint.

Have I ever seen it work out for a defense counsel to call his client to the stand at a motions hearing? Sure. But the potential disastrous consequences far outweigh the damage I have seen done to a case.

WHY RISK IT WHEN THERE IS A SAFE ALTERNATIVE?

The rules allow defense attorneys to supply sworn affidavits. *See* R.C.M. 905(h). This includes sworn affidavits from a military accused. Why subject a client to cross-examination when a controlled alternative is permitted under the rules?

CALLING THE CLIENT AT A MOTIONS HEARING

First, see above; do not do it. If you insist on calling the client to the stand during a motions hearing because you think you know better, prepare the client for the limited information he needs to provide on the motion and be ready to object to any questions that go beyond the scope of the purpose for which the client was called to testify. Some judges are far more permissive on what is meant by being within the scope of direct examination. This is a major reason that I strongly advise against ever calling a military accused to the stand for a motions hearing.

Even though most government counsels do not anticipate that a defense counsel will call the military accused to testify and rarely is the trial counsel prepared for any meaningful cross-examination, you cannot count on them not striking oil by accident. Do not rely on the potential for a government counsel not to be prepared; (don't call the client, but if you insist, make sure that you) prepare the client as though an experienced counsel will be ready.[31]

Recall that in drafting a defense pleading, the defense need not provide notice of their intent to call the military accused to testify. This, in part, may explain why most prosecutors do not anticipate the defense will call the military accused to testify about any matter.

---

[31] A separate book will cover my method for how to prepare a military accused to testify.

Jocelyn C. Stewart

Chapter 29

EXPERT EVIDENCE IN RULE 412 MOTIONS

EXPERT EVIDENCE

Realize that even if you have not yet employed an expert, you may need expert evidence to assist you in prevailing on a Rule 412 motion.

The most common example that I can provide of using expert evidence is expertise in forensic psychology or forensic psychiatry.

Expert evidence comes in many forms, and with access to scientific journals and other validated resources online or from general consultation with experts, you can provide the record with evidence that is separate and apart from expert testimony.

In addition to scientific journals and reliable secondary sources, often experts are willing to supply you with signed sworn affidavits to support your motion.

EXPERT TESTIMONY

Some fields are more complex than others. To make sure the judge understands your motion, it may be advisable to have the expert available for testimony. For example, the expert may provide your judge with additional information, may answer questions about connecting the dots, and may give a basis for their ruling. What may seem obvious to you from what you provided the Court may not be obvious to the judge. Having an expert available to testify during the Article 39(a) session likely will help you identify what are the Court's lingering doubts or hesitations to grant the relief you seek.

Sometimes practitioners will need actual expert testimony[32] to prevail on a Rule 412 motion. For instance, there may be evidence of prior risky sexual behavior of a

---

[32] I realize that during Rule 412 litigation, practitioners may also be working to compel expert assistance. Realize that often expert practitioners will assist you because they have an interest in being hired onto a case. If you ask nicely, that expert may also assist you in motions other than those which seek to compel their assistance.

Compelling expert assistance is particularly difficult under the military's standards. Absent demonstrating how the expert will assist you beyond what a Court may see as speculative, you are doomed to fail. The expert evidence will demonstrate in many cases the precise ways in which the expert will assist the defense. If you find yourself arguing to compel expert assistance and you have no documentation on which the Judge can make a factual predicate, expect to lose the motion.

In a recent motion to compel expert assistance, I crafted two separate draft affidavits for an expert in forensic document analysis. If you help the expert to help you by taking the time to write the draft affidavit for them, and you have taken the time to educate yourself on their field, that expert will be far more likely to assist you in reviewing, making any edits necessary, and then swearing to the information.

complaining witness. That evidence may coincide with behavior that is consistent with a borderline personality. Persons with borderline personality have issues with accurately reporting events, which bears on credibility. To have a shot at introducing evidence of prior sexual activity, the practitioner will need to use expertise to connect the dots. That becomes exceedingly difficult by citing only to journal articles. Believe me, I have tried.

I have found that no matter how clear the journal articles are or how well-crafted the affidavits, a Military Judge benefits from having the flexibility to probe the expert. An expert testifying on the motion will help them in obtaining any additional information for the factual predicate to support granting Rule 412 motions.

Never discount that despite your best efforts, there is a human element to motion's practice. No matter how prepared you are, there may be aspects of your factual predicate you could not anticipate would remain in question for the Court. So long as you are courteous to the expert, even if he is not yet hired onto the case, I have found that many are willing to make themselves available for telephonic testimony.

Part of what demonstrates professional courtesy to the expert is being respectful of her time. Ensure that you can control within reason the start time when she should expert to be called. An essential element of that courtesy extends to respectfully requesting to honor that estimation by communicating with opposing counsel and the Court about your preference of what order to litigate the motions. I have never seen a Military Judge who was unwilling to consider an expert's time when determining the order of march on motions litigation.

Realize that extending professional courtesies to a given expert will help you forge and keep strong relationships. In doing so, you will find that an expert is willing to extend that professional courtesy to you by providing no-cost help before they are formally hired onto the case.

Please bear in mind if you are seeking help before formal hiring, there are certain permutations of how to go about obtaining their help within the constraints of your attorney-client relationship before that expert is formally engaged. This is a large reason that employment of experts and motions to compel expert assistance and witnesses is a sufficiently complex topic that it warrants a separate book.

Jocelyn C. Stewart

Chapter 30

PREPARING FOR ORAL ARGUMENT – FACTS IN RULE 412 MOTION

IDENTIFY THE AREAS OF DISAGREEMENT

As Colonel (Ret) James Pohl repeatedly impressed upon counsel, judges know the law, but they need you to help them with the facts.

Identifying the areas of disagreement in the facts likely is the most crucial component of preparing to litigate any motion but can be particularly helpful in Rule 412 litigation. Most rules of Court outline that a response pleading needs to annotate which facts contained in the original filing that the responder does not agree with for purposes of the motion. Often those drafting the motion response pleadings overlook this requirement. Honestly, I wish more Military Judges would reinforce and mandate that counsel to abide by the provisions. It helps everyone to understand where, if at all, the facts are disputed and potentially will need further attention.

If it is unclear from your opponent's response brief, send a professional note to them via email to ask which facts from your fact section remain in dispute. You can even pick up the phone to call the counsel, though you should remember to try to memorialize any understanding in a subsequent email. By reaching agreement on some, most, or even all of the facts, you will identify which additional evidence, if any, you will need to provide at the motions hearing.

Whenever it is unclear which discrete paragraphs remain in disagreement, always take the extra step to seek clarification from the opposing counsel. The Military Judge greatly will appreciate your efforts to help streamline which areas he will need to make his findings. If counsel does not respond to a request for clarification for her position on the factual predicate, then the way that I attempt to streamline matters during the motion is to ask that the Military Judge please clarify from counsel which, if any, of the paragraphs of our facts that they dispute for purposes of the motion. In most circumstances, the judge will ask the opposing party if only for her own clarification as well.

Once you identify the paragraphs of the facts section that remain in dispute, make some notation on your working copy so that during the Article 39(a) you will be ready to present evidence on the factual predicate that is in dispute. The way you make these notes matters far less than using a system that reminds you of the area of disagreement. Recall that the paragraphs of your facts are organized as outlined in this book's predecessor and serve as your checklist for proving the required facts in your motion.

Often in Rule 412 motions litigation, government counsel inform the defense that they will "agree" to the sexual behavior or predisposition evidence solely for purposes of litigating the motion, not that they are agreeing the events are true for trial. Do not be caught off guard and do not assume these talismanic words mean you should veer off course from your litigation plan. Make sure you still call the alleging victim on the motion. I can easily recall more than a dozen occasions when the government attempted to "agree" to Rule 412 facts, but the complainant stated under oath those facts did not occur. Do not be led off the path of your plan.

Every counsel has different methods of preparing. However, technology can fail us. I recommend that you print a hard copy of your Rule 412 motion for your use to take with you to the hearing. You never know when your laptop may decide it requires an automatic update, or you forget your cord and the battery dies. If it can go wrong, it often does. The hard copy is your checklist if all digital notes become unavailable.

If you do not identify the areas of disagreement and you do not include evidence to support each required fact, the Military Judge will not be able to make findings of fact to rule in your favor. Remember that especially in Rule 412 litigation you have an uphill climb against this rule of exclusion. You will not make your job any easier by skirting your proof of the facts.

HELP THE JUDGE IN HIS FINDINGS OF FACT

The goal of a moving party in a motion is to provide substantiating evidence that will enable the Military Judge to make finding of facts that support your prayer for relief in the motion. In Rule 412 litigation, a finding of fact is a determination by the judge that there is evidence that a fact-finder certain could rely on to conclude a certain fact is true.

From my experience, it is always advantageous to try to confine a motions hearing to as little witness testimony as possible since witnesses can become nervous, forgetful, or may change their recollection. The most glaring exceptions are calling the complainant and calling the military accused.

Streamlining the factual predicate by using portions of the investigation including prior sworn statements and other affidavits you obtain for motions practice enables the Military Judge to decide the motion more easily in your favor. Often, the mistake of newer counsel is in their inability to recognize what facts are crucial for the Military Judge to receive in evidence upon which they will need to rely to make their findings.

Case law provides a roadmap as to what factors a trial court will or should use in deciding a given motion; the more cases you read on a given topic, the more you appreciate what facts you need to establish to obtain your requested remedy. The best time to go through that learning process is when you were drafting the motion. Sometimes the response brief may provide you with the opportunity to realize that you neglected a fact that will assist you in obtaining your desired relief.

In many jurisdictions, the judge will require you to submit proposed findings of fact and conclusions of law as a separate document from your motion. If you supplied the Court with proposed findings of fact, you may also use those as your blueprint for litigating the motion. Even if you did not, many if not most or all judges greatly appreciate a version of your motion in Word® format so that if your facts are supported by evidence, the Military Judge may adopt your facts in part or in totality in her ruling. Help your judge help you; organize the litigation from the already prepared Word® format of your facts. Using them as a checklist will help you highlight those areas where you need to focus proof in the hearing.

Jocelyn C. Stewart

Chapter 31

PREPARING FOR ORAL ARGUMENT – CASE LAW IN RULE 412 MOTIONS

ADDITIONAL LEGAL RESEARCH

Your legal research is not complete when you finish drafting the Rule 412 motion.

When you read opposing counsel's brief, as I previously discussed, it will assist you in clarifying any areas of factual disagreement. Reading the response brief will also identify additional areas of legal research to conduct.

Before you set foot in the courtroom to litigate the motion, you must read every case cited to in your opponent's response brief. Reading the response brief highlights (or should highlight) disputes involving interpretation of case law you cited to in your brief and will identify additional competing authority that your opponent is asking the judge to consider.

Often, government counsels cite military case law for a proposition that is wholly different from the holding of the case or even from dicta. Do not assume that the opposing counsel found case law to support their argument, and do not assume that their pleading accurately conveys the holding of a given case.

READ every case that opposing counsel uses in his pleading, and then be prepared to articulate during oral argument what each case conveys and what each case does not convey. If the case is properly cited, you will need to analyze how its facts are different from the facts of your case to distinguish it.

HOW TO READ A CASE

By read, I mean actually read each case mentioned in the reply brief, to include every case from those nasty string citations, for any purpose, and not just the head notes or summary. Chances are the counsel who drafted the response brief has not read beyond a parenthetical reference, and it is very likely they only copied and pasted the string citation from someone else's pleading they located on a shared drive.

Odds are strong that the counsel did not originally draft the majority of the pleading and almost never does counsel read the cases he cites. Since 2004, I have learned first-hand that counsel often cite to cases that do not stand for the proposition referenced. In prior cases, had I not bothered to read every word of the case the government cited to for a given proposition, I would have missed the dicta that helped me obtain relief.

While it might seem obvious, you should also Shepardize® each case mentioned in the opposing counsel's response brief. After ensuring the case was not overturned, check to see the date the case was decided. In many instances, the case may no longer be good law. For example, if the case was decided under an old version of Article 120, UCMJ, it may not have the same influence. Perhaps a landmark case makes the case no longer good law.

Remember that <u>Banks</u> was expressly repudiated in <u>Gaddis</u>. Did the opposing counsel cite to <u>Banks</u> and cases that cite to <u>Banks</u>? It matters.

DISCERN THE PROCEDURAL POSTURE

Likewise, figure out the procedural posture of the case. Was it a contested trial or was it a guilty plea? Did the trial defense counsel object at trial or is the appellant asking for the relief for the first time on appeal? It matters to what the case will mean to your situation.

In many briefs I have analyzed in preparation to litigate a motion, the opposing counsel cited without properly assessing its procedural posture. There is a grave difference in an appellate court determining that the error a trial judge committed in excluding Rule 412 evidence was harmless versus positive authority that the evidence the defense seeks to introduce is legally permissible and on solid ground.

Read the case.

The procedural posture matters.

Harmless error does not translate to legally excludable, yet frequently government attorneys do not always translate the difference in their briefs.

BE PREPARED TO DISTINGUISH CASES OFFERED BY OPPOSING COUNSEL

No matter what the opposing counsel says about a given case, do not trust it until you have reviewed it and decided that it says what they claim it says.

In Chapter 34, I will discuss methods of sifting through cases cited to by your opposition, but for now, realize that when you are preparing for oral argument that you will need to pay close attention to the precedent established by the case, if any, the procedural posture of the case, whether the cited to proposition is a holding or is dicta, and whether the case is factually distinguishable from your own.

Chapter 32

ORAL ARGUMENT ETIQUETTE IN RULE 412 MOTIONS

DECORUM ABOVE ALL ELSE

Counsel will lose credibility quickly if they do not exhibit professionalism during motions practice. Beyond preparation, counsel that practice in a military setting do well to remember customs and courtesies and must NEVER directly address (speak to) an opposing counsel on the record.

ADDRESS ONLY THE MILITARY JUDGE

All comments must be directed toward the Military Judge. If there is an issue for which clarification must be sought from the opposing counsel on the record, you must direct the colloquy to the Military Judge. For instance, instead of addressing the opposing counsel, speak to the judge and ask that the judge seek clarification from the opposing counsel. An effective manner to maintain such decorum would be to state to the judge, "Your Honor, we respectfully invite the government counsel to provide clarification as to whether or not they are opposing the motion in its entirety or whether they are merely opposing the portion of our motion regarding the sexual items on display in the complaining witness' barracks room."

Never turn to the opposing counsel to speak to them on the record. It is wholly inappropriate and may (rightfully) damage your credibility with the Court.

DO NOT GO LOW

Even when you believe the opposing counsel is flat wrong on her interpretation of case law, there is a distinguished manner to accomplish the noted differences in your position. For instance, instead of saying "CPT X is wrong" state "defense respectfully disagrees with the government counsel's interpretation."

There is no need to insult an opposing attorney no matter how wrong you believe your opposing counsel's viewpoint. Rudeness to the other side will not win you any points with the Court. A Military Judge would much rather focus on officiating the law than on dealing with refereeing the snide comments of counsel.

The judge is far more likely to side with you if she is not charged with sifting through rhetoric. Make the litigation about the case, not about any personal need to look superior. Any effort on the part of opposing counsel to invite you into such side issues should be ignored as useless noise. You will find the judge more open to your arguments if you remain calm and even tempered.

Jocelyn C. Stewart

Chapter 33

ORGANIZATION OF ORAL ARGUMENT IN RULE 412 MOTIONS

WRITTEN MOTION ORGANIZES THE ARGUMENT

By following the P-SAC® method laid out in ***Shaping the Battle: How To Write Motions In Military Practice***®, you have already set yourself up for proper organization of your oral argument.

Recall the argument format for military writing in motions practice that conveys the clearest message is one that follows the following "P-SAC" format:

- **P**remise
- **S**upport
- **A**nalysis
- **C**onclusion

PREMISE

What do you need the Military Judge to conclude in order to grant you the requested remedy?

SUPPORT

What are the rules and cases that support your premise?

ANALYSIS

How do the FACTS of your case fit the support you provide to lead the reader to conclude that your premise is valid?

CONCLUSION

Applying the FACTS of your case to the supporting rules and cases, your premise is the right result.

HEADERS

When drafting your motion, recall that no matter how well reasoned your legal argument / analysis section, your arguments are bound to be lost in a sea of paragraphs unless they are organized around headers. By making each header a declaration of the legal arguments you are crafting both an organized and a persuasive framework, you are also preparing yourself for oral argument.

When you choose your headers and the framework of your argument based on the standard set out in the law for the particular area of your motion, you also choose the organization of your oral argument. The declarative statements are your argument, so that if you become nervous or anxious during argument, you can default to those headers to remind you the key parts of your argument.

Chapter 34

SUBSTANCE OF ORAL ARGUMENT IN RULE 412 MOTION LITIGATION

DO NOT REGURGITATE YOUR WRITTEN PLEADING

Military Judges are literate; they are able to read your written motion. As the moving party, do not use oral argument *only* to repeat what you establish in your pleading.

PREPARE A BRIEF RESTATEMENT TO FOCUS IN ON THE ISSUES

In most military courtrooms, the judge has already focused on the issue or issues that must be resolved before ruling. Seldom will you have much opportunity to give more than a brief restatement of your position.

Do your homework on the judge. Maybe you have litigated in front of this judge before and know whether you will be peppered with questions or whether you should be prepared for a brief speech.

PURPOSE OF ORAL ARGUMENT

Oral argument should be used to address any holes in your pleading, to address the arguments of opposing counsel, and to distinguish case law the opponent cited to in the opposing counsel's response pleading. When the Military Judge directs you to make oral argument, professionally state that you wish to use this opportunity to address opposing counsel's arguments in their response pleading and / or cases the opposing counsel cited to in her response brief.

HOW TO DISTINGUISH CASE LAW

After outlining how you plan to proceed, address the cases you need to distinguish one by one. State to the Military Judge, words to the effect that "unlike the holding for which the government / defense cited the case of *U.S. v.* _____ for in their brief, *U.S. v.* _____ stands for the proposition that Y."

Be respectful but be firm and clear with the Military Judge that you have "done your homework" and have taken the time to read the cases the opposing counsel used in their brief.

DISTINGUISHING CASE LAW PROCEDURALLY

Often, the opposing counsel will cite a given case that is procedurally distinguishable. For instance, on appeal if the appellant pleaded guilty, the issue may be waived. If the appellant did not raise the issue at trial, the standard of review will be more difficult to reach. In cases where defense counsel did not raise an issue at trial even if not waived, the review standard also is far more difficult to reach. Cases with a different procedural posture are easy to distinguish.

Beware of cases that opposing counsel cites to that were guilty pleas. In guilty pleas, most issues that were not objected to or raised at the trial level are waived. Your motion necessarily raises the issue. As long as you explicitly cite to each legal basis for raising the motion, the issue will not be waived.

ANALYZING THE TYPE OF ERROR

Be leery of cases the government cites to where the appellate court found error but determined that the error was harmless beyond a reasonable doubt. I cannot underemphasize how often this happens in reply-briefs, so please watch out for it. Ordinarily an appellate court will decide that the error was harmless because the other evidence in the case was overwhelming in terms of guilt, usually in the form of military accused's admissions or outright confession. Ensure that the Military Judge understands that the same issue constituted error. Error matters most.

On occasion the opposing attorneys plant seeds of support in its motion for your position that you may not have found. Do not underestimate the opposing attorney's misread of any given case. Often counsel shortcut response pleadings and rely on headnotes that seldom convey a given case's holding or its context and significance.

Another common occurrence is that counsel cite to a given case as authority that the Military Judge *should* exclude a given piece of Rule 412 evidence, but upon additional review, the case was not in a parallel posture to lend credibility to the counsel's argument. For example, it is not helpful to cite to a case that actually stands for the proposition that the trial judge committed error by excluding the Rule 412 evidence, but that the error was harmless because of "overwhelming" evidence in the case. This is usually code for the accused person confessed or made significant admissions.

Realize that any determination that error occurred is useful, but make sure that when you address the analysis on the case, you highlight the distinction from the facts of the reported case when compared with your own. The most significant distinction is often that in the reported case, the prosecution had accused admissions or an outright confession. If your client did not make admissions, be ready to highlight this material difference.

Remember that because of the fundamental requirement of candor to the tribunal,[33] you have a duty to report case law that cuts against your position. If a case tends to support a different result than the one for which you seek judicial relief, comb through the facts and be prepared to distinguish the facts from your own case's factual circumstances in the analysis portion of your pleading. Do not ignore the case; do not rely on opposing counsel or the Military Judge to cite it for you. It exists, and you must address it.

PREPARING NOTES FOR ORAL ARGUMENT

When I prepare notes for oral argument, my first section relates to addressing the cases that opposing counsel cited to that I did not in my pleading. I provide a two or three sentence synopsis for the facts of the appellate case. Demonstrating your preparation is important to the judge. In addition to my notes on each case, I copy a PDF version into my oral argument computer file of each of the cases I am addressing during oral argument so that I have it ready if I need to look at it during oral argument to field

---

[33] American Bar Association Model Rule 3.3.

a judicial question. For any case that is fundamental to the motion, I will the actual cases printed out and handy in case my synopsis is not enough for me to recall its significance.

The second section of my notes for oral argument highlights any areas in the opposing attorney's response brief that cited to facts that are inconsistent with the proof supplied. Sometimes counsel confuse cases and may argue the wrong set of facts in their brief. Be ready to guide the judge to the support for your factual predicate.

The third section of my prepared notes during oral argument is for the questions that I anticipate the judge will ask. Details on anticipating those questions follows in Chapter 35.

The last section of my preparation notes for oral argument are for reacting to the oral argument of my opponent. There are gems to mine in what opposing counsel says during oral argument. Do not be distracted by preparing for the next motion to argue in the lineup after you make your argument. Prepare the other aspects of your oral argument before oral argument happens. Then, be ready to listen to the opposing argument and be prepared to answer it.

Jocelyn C. Stewart

Chapter 35

ANTICIPATING QUESTIONS BY THE MILITARY JUDGE

BE THOUGHTFUL

At the outset anticipate that the Court will ask you on whom the burden rests and what is the standard of proof is for the party on whom the burden rests.

Hint: Rule 412 motions do not carry the same burden as other motions. Remember that in Rule 412 litigation, the proponent need not prove that the evidence being offered is true by a preponderance of evidence; it is not for the judge to decide as to whether the judge believes the truth of the evidence being offered, although it would not hurt if your evidence was convincing.

Anticipate that the Court will ask you for the seminal case if one exists in this area of the law. If you followed Part I of this book, your motion should already have covered the seminal case.

DANGER POINTS

Ask yourself what is the aspect of your motion that you are most afraid of, and you will identity the danger point of the motion's litigation. Here is where you should anticipate that the Court will direct its questions. Often the response pleading does not identify the weakness of your motion, but you should expect that the judge will.

DEVIL'S ADVOCATE

Once you have identified the areas of weakness, or your danger points, confer with a co-counsel or even practice with yourself. How will you answer the Court's questions? Anticipate them so that you will be ready for them. In Rule 412 litigation, the danger points or weakness in nearly every case is surviving any required balancing test.

WORRY LESS ABOUT THE RESPONSE PLEADING IN ANTICIPATING JUDICIAL QUESTIONS

In a 2021 oral argument on a Rule 412 motion, I anticipated that the Military Judge would be focused in on why I needed the extra "detail" that the alleging victim's source of employment had been sexual activity on OnlyFans®. I anticipated he would want to "split the baby", and he did. The judge's questions centered on whether the only bias I needed in evidence was that my client had cut off the alleging victim's ability to earn additional income? His questions went to the crux of whether that was the only salient fact I needed to raise the putative victim's motive to fabricate. From his questions, I had correctly anticipated that he would not be inclined to allow the fact that it had been employment of a sexual nature that my client stopped. I anticipated accurately that would be the focus for the judge.

Because I had already anticipated this issue, I had prepared and answered in terms of <u>Collier</u>.[34] Despite its ruling, the dicta of Collier is the friend to the defense.

In <u>Collier</u>, a trial judge refused to permit cross-examination about a same-gender relationship that ended in a bitter split. The trial judge only permitted introduction of a friendship that was no longer in existence and ruled that was essentially enough to get after the bias. On appeal, while the appellate court did not reverse the conviction, <u>Collier</u> provides support for the idea that the defense had failed to connect the dots for why the same-gender relationship's ruin was more probative or material than a friendship that ended.

I weaved my intended answer back to the judge and explained factually the motivation to fabricate ran much deeper than a loss of money. She was enraged because my

---

[34] 67 M.J. (C.A.A.F. 2009)

client had shamed her in a judgmental and puritanical way. Merely being told that he caused her to lose money was one thing, but that it was coupled with a moral judgment made the bias that much deeper.

Despite my anticipation and crafted argument (and my present day belief that I was correct) in that case, the judge disallowed that the source of income was OnlyFans®. He also excluded that my client had shamed the alleging victim for having lingerie and toys for sexual stimulation in plain view during a barracks inspection. We were only able to mention that my client had yelled at her and corrected her. The client was acquitted anyway.

As an am important aside, during that same trial, the panel wanted to know the source of income and the panel wanted to know what specifically my client had corrected in the alleging victim's barracks room. They posed this question of my client while he was testifying during merits. When asked by the same Military Judge who had excluded the evidence how we should handle the situation, I proposed an instruction[35] to the panel that made them not knowing far more valuable to my case than providing them with an answer. The panel intimated that whatever it was,

[35] The next feature in the Shaping the Battlefield Series will be on Panel Selection and Instructions in Military Practice.

it was so bad that the judge would not let them hear it. That said more than the evidence could have.

Chapter 36

SUPPLEMENTING THE MOTION

UH-OH!... DON'T WORRY

During oral argument, it should become clear if the Military Judge believes that you are missing a key component of proof required to meet your burden on the motion. All is not lost.

Many if not most or all judges will permit you to supplement the motion after oral argument. Respectfully state that you realize that you have additional material to provide the judge to help in their consideration of the issue. Sometimes this will necessitate writing a separate pleading to address the discrete issue, and sometimes it will only be to send additional matters to the judge and opposing counsel.

If you decide you need to supplement the motion, anticipate that the judge will afford opposing counsel a chance to supplement as well. Do not rely on supplementing when preparing to write the motion or to litigate it, but then realize that it would be better to ask to supplement than to lose your motion for failing to establish proof.

Even if you do not realize in the moments of oral argument that supplementing is necessary, a contrary ruling may prompt you to see and appreciate where your evidence was lacking. In these cases, do not be afraid to seek additional evidence and to revisit the motion in a motion for reconsideration.

Realize that motions to reconsider in Rule 412 motions can be more common than in other litigation because Rule 412 is a rule of exclusion. If you realize that you failed to connect the dots of materiality, for example, then do not shy away from filing a pleading that more plainly outlines why the evidence is so critical. Reconsideration can also be helpful if you realize that you did not establish sufficiently that the Rule 412 evidence impacted the client's thoughts and beliefs about consent.

Chapter 37

DIGESTING THE RULING IN RULE 412 MOTIONS

DIGEST, ASSESS & (MAYBE) REATTACK

When the Rule 412 ruling comes in, remain calm if the judge ruled against you. There is a great deal to discern from a ruling, which may give you a clear path to seeking reconsideration.

Look to the ruling to see where the judge believes you fell short, and then determine if there is a way to overcome. In most circumstances, the judge will outline which prong of a test she believes you failed to meet. Ask yourself whether the failure is one of factual predicate.

If it is a factual matter, assess whether you are able to conduct further investigation to obtain additional proof for the Court. Additional facts will be the area where you have the best chance to obtain relief on reconsideration.

If there are errors in the findings of fact, you will need to make a tactical calculation about the impact of this motion on your case. What will be the standard of review on appeal? If the judge made findings that were clearly erroneous, the standard of review will be different.

WIN IT AT THE TRIAL LEVEL

I am aware that some attorneys might encourage others to let an issue alone so that on appeal, there can be error to point to. Sometimes, these are known in common parlance as "appellate bombs." I generally take the opposite stance. In most instances, my advice is to try to "win" the issue at trial, rather than to risk relying on relief on appeal.

There are tactful ways to address erroneous findings in a reconsideration. Avoid citing to the same source and highlighting the mistake. Remember that judges are human, which impacts you two-fold. First, they are capable of mistakes. But secondly, and perhaps most importantly, this person will be the one to rule on this potential reconsideration and a litany of other rulings, has a human ego. Realize that a human is far more likely to consider issues favorably for your client if you did not opt to alienate or embarrass them needlessly.

WRONG STANDARD

If the judge applied the wrong legal standard, you should seek reconsideration. Perhaps the Court applied the wrong balancing test, and the Rule 412 evidence is being offered under the constitutional exception.

KEEP THE RULING IN YOUR WAR CHEST

Whether you were successful or not, and even if reconsideration is not appropriate in your Rule 412 case, every judicial ruling is a treasure to keep in your war chest moving forward. The ruling provides you with insight into the judge's logic. Do not merely scroll to the bottom line ruling and then discard it. There are gems to be had for future use. In many instances, a judge's ruling demonstrated to me that the judge had made a connection about some of the facts that I had not. Remember that at their foundation, the Military Judge is another military officer who is analyzing your case and often without some of the blinders you may carry. In some ways, they are more like your potential panel member than you are given their age, experience, and service. Do not discard a ruling as the end of things.

Use a prior Rule 412 motion ruling as framework in future motions that you file to the Court. When I have litigated a Rule 412 motion on a prior occasion in front of the same judge, I go back to the prior ruling and use it as a checklist before I file my new Rule 412 motion. Why would I not? Do not be so conceited as to think you cannot benefit from the framework the judge provided you. The Court's ruling on Rule 412 motions provides substantial insight into what cases the Court sees as most important, and which facts or type of facts weigh most heavily on him.

Using a prior ruling as a checklist is helpful but avoid the temptation to simply copying the prior judicial ruling as your own motion.

Jocelyn C. Stewart

Chapter 38

RECONSIDERATION IN RULE 412 MOTIONS

ASKING FOR RECONSIDERATION

As long as there is a reason for it, do not be afraid to ask for reconsideration. However, do not ask for reconsideration only because you did not like the outcome.

FORMS OF RECONSIDERATION

Reconsideration can come in the form of a formal pleading. Reconsideration can also come from an oral motion made when the issue is ripe for the judge to reconsider a prior ruling during trial.

FORMAL PLEADING

When something new arises that gives you a justification to ask for reconsideration, file a request that the judge reconsider coupled with a request to file out of time. Ensure that you have additional evidence on which you are making your request. Supply an additional affidavit or a supplemental one from a witness who had given previous evidence.

ORAL MOTION

In some instances, the realization or impetus for reconsideration may not become apparent until sometime during trial. Do not be afraid to ask for reconsideration. Be respectful and do not make claims that are unsupported, but if there is a legitimate reason to ask for reconsideration, be sure to ask. Even if you lose, there is a record at stake.

In an Air Force trial in Georgia several years ago, the judge had prohibited introduction that the alleging victim had suffered a prior sexual assault. Our theory had always been that the emotional strain of that prior incident was a factor that weighed on the complainant in how she (incorrectly) viewed her encounter with our client. We felt strongly that the panel needed to assess her credibility through that lens in our trial. We were not trying to get into any details, but it mattered to our case that she was a prior victim. After Rule 412 litigation, the military judge ruled any mention of her prior assault was excluded.

Based on the alleging victim's demeanor and presentation, during the trial a panel member question asked whether the complainant had ever suffered prior trauma. During at least three separate closed Article 39(a) sessions, I asked the Court to reconsider its prior ruling. In the last and triumphant one, I stated that without waiving the issue on appeal, I was asking that the Court at least allow us to elicit that the alleged victim had suffered a prior *trauma* without mentioning that it had been a prior sexual trauma. The military judge finally ruled in our favor. The client was acquitted of non-consensual sexual misconduct, and I am firmly convinced this reconsidered ruling played a substantial role in our success.

Chapter 39

WRAP-UP IN MOTIONS LITIGATION

The most significant misstep that I see in litigating Rule 412 motions is a failure to adequately prepare. Preparation for oral argument begins with drafting the motion. Systematic drafting ensures systematic preparation. Oral argument is only a part of litigation. What is most important in litigating Rule 412 motions is digestion of the body of caselaw and its applicability to the facts of your case. If you shortcut the process, you are shortcutting your case, which in turn can mean disaster for your client.

I am excited to finish this book. I struggled to write it far more than any other project I have undertaken. That makes its completion all the more satisfying.

Rule 412 litigation shapes Article 120 cases more than any other type of motion I can imagine. My goals will always be the betterment of military courts-martial practice. With so much of military prosecutions focusing on Article 120 cases, I cannot imagine shaping the practice in any more significant way than to complete this book and provide it for use by our practitioners on the front lines of impacting service members lives. Thank you for the opportunity to contribute.

TOP TEN PRACTICE POINTS IN RULE 412 LITIGATION

1. **Read the Response Brief**. No really. Read it.

2. **Start with** their **Facts** section. Ascertain what facts are in dispute; this will streamline the process and will make your judge very happy. It will also help focus you on what areas you will need to focus on where certain facts are disputed.

3. **Check their string citations** and what they purport to stand for; chances are the case does not stand for the proposition that it purports to. **Download *and read* all cases** the opposing counsel cites to. Do NOT TRUST that the response brief's author did more than find some string citations or read a few headnotes. Do the work. You'll find gems.

4. **Identify whether the motion's drafter is citing to cases with the right procedural posture**. Not all citations in a motion are accurate. Do your work and verify each one.

5. If the opposing counsel cites to a case that you did not, but that does offer some actual support for their position, **read, and brief the facts in a word doc** that you have readily available for oral argument. Then distinguish the facts of that case from the facts of your case.

6. Supplement, supplement, supplement. Consider obtaining a supporting affidavit to provide as additional evidence.

7. In most motions' litigation, my advice is to **avoid calling most witnesses if at all possible**. The most notable exception to this rule is **calling the complainant in Rule 412 motions.**

8. During oral argument, **Avoid Repeating your Pleading**. If the opposing counsel's reply brief offers nothing in the way of law or other substantive support to oppose your motion, state it. Don't be afraid to "rest on the motion, subject to any questions of the Court."

9. Use Oral Argument to do Something New: Respond to the reply brief, provide additional cases you found after submission of your original pleading, argue additional facts you supplemented, and offer to answer questions of the Court.

10. Put on evidence that is necessary to the resolution of any matter. For Rule 412 motions especially, do not assume that Judge will connect the dots. Do it for him.

Jocelyn C. Stewart

About The Author

Jocelyn C. Stewart attended Louisiana's designated honors college, the Louisiana Scholars' College, in Natchitoches, Louisiana from 1996 to 2000. After being awarded a three-year ROTC cadet scholarship and a scholarship from the honors college, she graduated first in her class of honors students, summa cum laude, and "with highest distinction." Ms. Stewart commissioned Distinguished Military Graduate from Northwestern State University's Demon Battalion as a Second Lieutenant in Military Intelligence. She was selected to attend law school as an educational delay student from 2000 – 2003.

Since Jocelyn graduated law school from the Paul M. Hebert Law Center at Louisiana State University in 2003, she has exclusively practiced military law.

An active duty Army JAG Corps from early 2004 until late 2012, Ms. Stewart worked in uniform for more than seven years in court-martial practice. Her initial assignment was as a legal assistance attorney and a Part Time Military Magistrate where she reviewed law enforcement applications for search and/or seizure authorizations and command decisions to place servicemembers in pretrial confinement pending their courts-martial.

In 2005, Jocelyn became a "trial counsel," serving as the command's legal advisor in military justice matters and prosecuting any of its Soldiers facing military criminal trial. After serving as trial counsel for just over 2 years, Jocelyn fought to stay in the courtroom.

In 2007, she took over as a trial defense counsel at Wiesbaden Army Air Field (WAAF), Germany. In 2008, she moved to the then busiest court-martial jurisdiction of any of the services: Fort Hood, Texas. From 2008 until 2010, Jocelyn defended multiple high profile cases including premeditated murder, the kidnapping of a two-day old infant from the base hospital, and a Soldier who staged his own "kidnapping" by members of the Mexican cartel.

In 2012, Jocelyn was hand-selected to stay in a courtroom advocate role and fill one of the first Special Victim Prosecutor positions. Responsible for the Midwest region, Ms. Stewart handled all "special victim" cases for Fort Riley, Fort Leavenworth, and Fort McCoy. A "special victim" case is defined as any allegation of adult sexual assault by a civilian or a military member, any allegation of child sexual abuse or molestation, all allegations of intimate partner violence, and those with evidence of digital exploitation of children. For two years, Jocelyn headed and oversaw the prosecutions of more than sixty special victim cases.

As a Special Victim Prosecutor (SVP), she achieved the only two non-homicide life without parole sentences in the Army's history. In addition to the supportive and instructive role she played in the courtroom for the junior trial counsel in her region, Ms. Stewart's job as SVP included extensive travel to teach prosecution techniques to trial counsel of all branches of service.

In 2012, Jocelyn left active service to pursue a practice dedicated to defending servicemembers worldwide, with particular emphasis in sexual assault investigations and sexual assault court-martial defense. As a civilian court-martial specialist, she has represented clients OCONUS and all over CONUS, from the west coast to the east coast, the gulf coast, and the Midwest. Her clients serve the Army, the Air Force, the Navy, the Marine Corps, and even the Coast Guard.

After a three-year break in service to focus on building her civilian court-martial practice, in late September 2015,

Jocelyn assessed into the reserve component of the Army JAGC. She first served for more than fifteen months as the Anchorage Team Leader of the 6th Legal Operations Detachment (LOD). She also served as the 6th LOD's Public Affairs Officer.

In January 2017, Jocelyn joined the adjunct faculty of the Criminal law department of The Judge Advocate Generals Legal Center and School. In recent memory, Jocelyn was the first defense attorney selected to join the faculty, which historically has consisted of federal prosecutors. After competing with dozens of applicants and undergoing a thorough screening and interview process, Jocelyn was invited to join the faculty. During her three-year adjunct faculty tenure, Jocelyn taught hundreds of RC / AC Judge Advocates about UCMJ criminal case updates during the 2018 Western Reserve On-Site. She has also taught nearly one hundred trial counsel and defense counsel at the Intermediate Trial Advocate Course (ITAC) in several iterations of the course.

Lastly, Jocelyn taught two separate iterations of TJAGLCS LLM graduate students about military motions practice and graded multiple LLM graduate course final papers. Ms. Stewart's tenure on the Criminal law faculty ended effective January 9, 2020.

In 2021 and 2022, Jocelyn was invited to instruct dozens of uniformed defense counsel at a Defense Counsel Assistance Program (DCAP) conference held in Arizona. She teaches for several days on topics including motions writing and litigation in military practice. She also teaches counsel about resilience in litigation, how to work with civilian counsel, and overall strategies in serving as a defense counsel in military court-martial practice.

In 2021, Jocelyn also promoted to the rank of Lieutenant Colonel in the Reserve Component. She is expected to retire from the reserve component in 2023.

Ms. Stewart can be reached through her firm at:
www.UCMJ-Defender.com

Jocelyn C. Stewart

About The Firm

Unlike many of her competitors, the Law Office of Jocelyn C Stewart, Corp. focuses from day one at running a collateral investigation to understand the state of the government's evidence and to attempt to find evidence the government has yet to uncover. The firm's ability to make the most significant impact is linked to how early clients find us. Investigation is the most important aspect of any case preparation and in many cases, is the best opportunity to stop a case before the preferral stage. However tempted a service member may be to "wait and see what happens" most often it is a mistake to wait.

Don't wait; contact the firm today.

Law Office of JOCELYN C. STEWART
1201 Pacific Avenue
Suite 600
Tacoma, WA 98402
Office: 253-212-9582
Toll Free: (888) 469-2260
FAX: 1-888-252-0928
www.ucmj-defender.com

www.ingramcontent.com/pod-product-compliance
Lightning Source LLC
Chambersburg PA
CBHW071630200326
41519CB00012BA/2237